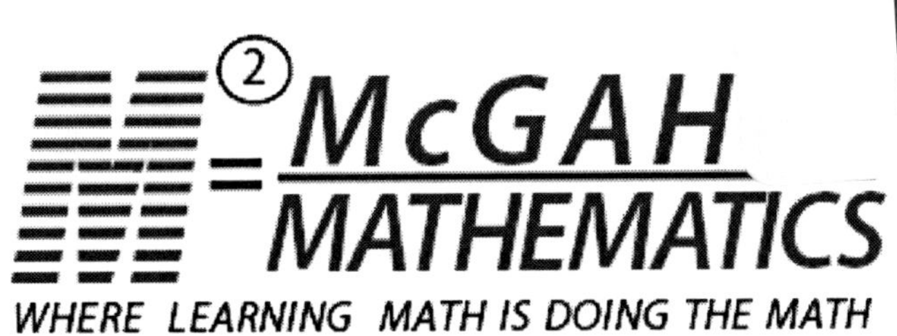

ALGEBRA WORKBOOK

Ben McGahee, M.A.

Professional Mathematics Tutor

And Mathematics Specialist

TOPICS

- Linear Equations
- Absolute Value Equations
- Linear and Absolute Value Inequalities
- Graphing Points in the Plane/Slope-Intercept Form
- Parallel and Perpendicular Lines/Systems of Linear Equations
- Polynomials/Adding and Subtracting Polynomials
- Multiplying Polynomials
- Dividing Polynomials
- Factoring Polynomials
- Rational Expressions
- Rational Equations
- Radical Expressions Without Variables
- Radical Expressions With Variables
- Complex Numbers
- Quadratic Equations
- Quadratic Inequalities
- Radical Equations
- Rational Inequalities
- Midpoint and Distance Formulas
- Properties of Logarithms
- Exponential and Logarithmic Equations

- Relations and Functions
- Non-Linear Functions
- Composite and Inverse Functions
- Matrices, Determinants, and Cramer's Rule

LINEAR EQUATIONS

QUESTIONS **ANSWERS**

Solve each equation.

1. $x+5=21$ ____________
2. $14-y=33$ ____________
3. $2a+10=-6$ ____________
4. $7z-18=11z-42$ ____________
5. $\frac{m}{5}+\frac{3m}{10}=\frac{1}{25}$ ____________
6. $-9(p+4)=2p-15+p-13$ ____________
7. $\frac{t-2}{8}=\frac{2t+1}{11}$ ____________
8. $\frac{1}{5}(20w+15)=\frac{3}{4}(-16-24w)$ ____________
9. $3(b-18)+4b=8b-(b+50)$ ____________
10. $6n-9+2n+5=4(2n-1)$ ____________

Solve each equation for the given variable.

11. $ax=b;x$ ____________________

12. $\frac{y}{p}=k;y$ ____________________

13. $m+nt=z;t$ ____________________

14. $q(v+b)=f;v$ ____________________

15. $\frac{u}{r} = \frac{u+t}{r-s}; u$ ____________________

Solve each word problem.

16. A number is four more than twice that number. Find the number.

____

17. The total cost of a t-shirt with 8% sales tax is $13.50. What is the original price of the t-shirt?

____

18. The perimeter of a rectangular garden is 220 feet. If the length of the garden is 80 feet, what is the width of the garden?

____

19. Two cars travel in opposite directions from the same location at 3 p.m. One car travels east at 45 miles per hour while the other car travels west at 60 miles per hour. At what time are the two cars 525 miles apart?

____

CHALLENGE

20. A baker can make c cookies in h hours. How many days d can the baker make k cookies?

____

21. A son weighs w pounds and his father weighs W pounds. They plan to sit on a balanced seesaw. The distance between the father and son is L. If the son is sitting a distance of x units from the fulcrum and the seesaw is balanced, then what is x in terms of w, W, and L?

    Hint: Multiply the weight of the person times the distance from the fulcrum for both the father and son.

____

SOLUTIONS

1. $x = 16$
2. $y = -19$
3. $a = -8$
4. $z = 6$
5. $m = \frac{2}{25}$
6. $p = -\frac{2}{3}$
7. $t = -6$
8. $w = -\frac{15}{52}$
9. no solution
10. all real numbers
11. $x = \frac{b}{a}$
12. $y = kp$
13. $t = \frac{z - m}{n}$
14. $v = \frac{f - bq}{q}$
15. $u = -\frac{rt}{s}$
16. -4
17. $12.50
18. 30 feet
19. 8 p.m.
20. $d = \frac{hk}{24c}$
21. $x = \frac{LW}{w + W}$

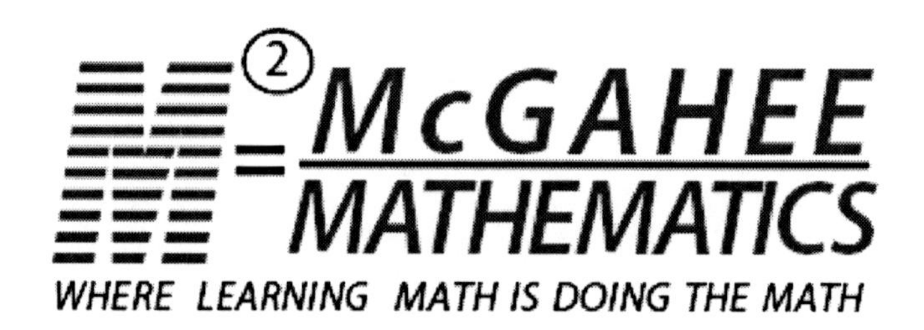

ABSOLUTE VALUE EQUATIONS

QUESTIONS **ANSWERS**

Solve each absolute value equation.

1. $|x| = 2$ ____________
2. $|y-3| = 9$ ____________
3. $|4z| + 17 = 53$ ____________
4. $|2a+1| = |8-6a|$ ____________
5. $|n| = -10$ ____________
6. $\left|-\frac{3}{4}t + 15\right| = 0$ ____________
7. $16|w-24| = 80$ ____________
8. $|m| + 31 = |2m| - 41$ ____________
9. $|15(p-2)| = |30p|$ ____________
10. $\left|\frac{7r+4}{3}\right| = 20$ ____________

Solve each absolute value equation for the given variable.

11. $|ax+b| = c$; x ____________
12. $|z-d| = |d-z|$; z ____________

Solve each word problem.

13. The distance between a number and three is 8 units. Find the number(s).

____

14. The distance between three times a number and negative four is ten. Find the number(s).

____

15. The distance between a number and five is the same as the distance between twice a number and fifteen. Find the number(s).

____

True or False: If the statement is false, explain why the statement is false.

16. $|ab| = |a|\,|b|$ for any real numbers a and b. True OR False

17. $|a + b| = |a| + |b|$ for any real numbers a and b. True OR False

18. If $|ax + b| = d$ and $d = 0$, then $x = -\frac{b}{a}$. True OR False

19. The distance between a number and zero can be negative. True OR False

CHALLENGE

20. The distance between x and a is d. The distance between x and b is $2d$. Find the largest distance between a and b.

____

21. Let p and q be two positive numbers in that order. Find the smallest number between p and q such that the distance between that number and q is twice the distance between that number and p.

____

SOLUTIONS

1. $x = 2, -2$
2. $y = 12, -6$
3. $z = 9, -9$
4. $a = \frac{7}{8}, \frac{9}{4}$
5. no solution
6. $t = 20$
7. $w = 29, 19$
8. $m = 72, -72$
9. $p = -2, \frac{2}{3}$
10. $r = 8, -\frac{64}{7}$
11. $x = \frac{c-b}{a}, \frac{-c-b}{a}$
12. z is any real number
13. 11, -5
14. 2, $-\frac{14}{3}$
15. $x = 10, \frac{20}{3}$
16. True.
17. False. Let $a = 1$, and $b = -1$. Then $|1 - 1| = |1| + |-1|$, which implies that $0 \neq 2$. Other similar examples will suffice.
18. True.
19. False. The absolute value of a number is the distance that number is from zero. Distance can only be a non-negative value.
20. $3d$
21. $\frac{2p+q}{3}$

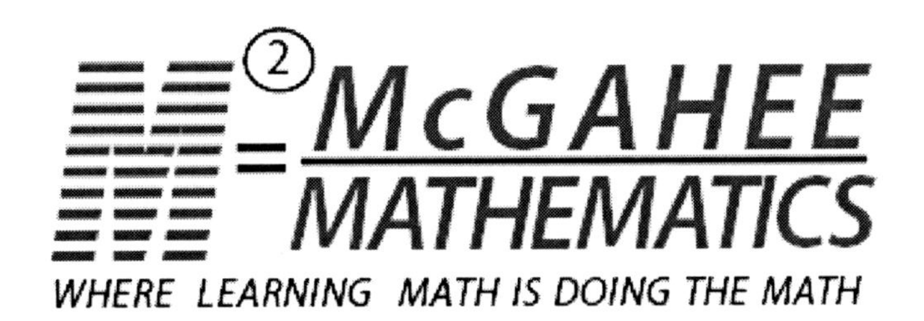

LINEAR INEQUALITIES

QUESTIONS **ANSWERS**

Solve each linear inequality. State the solution in interval notation.

1. $x + 7 > 25$ ____________
2. $3y \leq 108$ ____________
3. $-2z + 9 \geq -11$ ____________
4. $6(a - 4) + 12 < -5a + 10$ ____________
5. $\dfrac{t-1}{2} \geq t + 1$ ____________
6. $\dfrac{1}{7}(14k - 21) \leq \dfrac{1}{5}(10k + 60)$ ____________
7. $\dfrac{t}{8} + \dfrac{5}{12} < \dfrac{3t}{4} - \dfrac{1}{3}$ ____________

Solve each word problem.

8. If $ax - b \leq c$ and $a < 0$, then what is x?

____

9. The perimeter of a rectangular field can be no more than 800 meters. If the width is 20 meters less than the length, then what is the maximum area of the field?

____

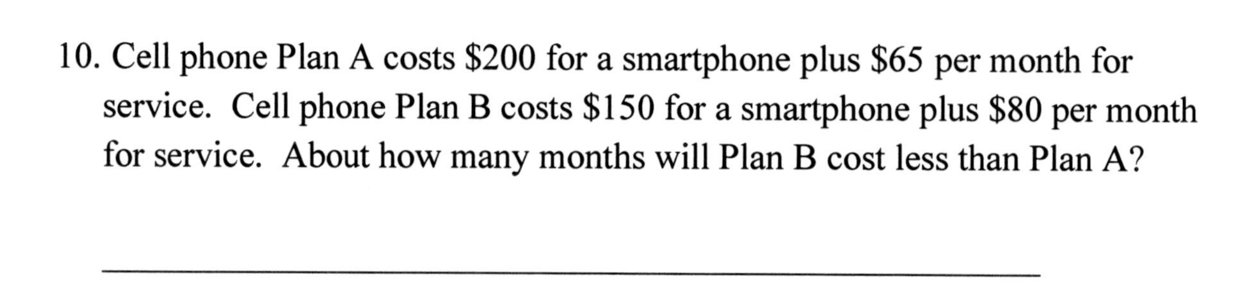

10. Cell phone Plan A costs \$200 for a smartphone plus \$65 per month for service. Cell phone Plan B costs \$150 for a smartphone plus \$80 per month for service. About how many months will Plan B cost less than Plan A?

ABSOLUTE VALUE INEQUALITIES

Solve each absolute value inequality. State the solution in interval notation.

11. $|x+7|<13$
12. $|2y-5|\geq 9$
13. $|3-4z|>-21$
14. $|8t|+10\geq 42$
15. $|6w|-15\leq -33$

Solve each absolute value inequality for the given variable. State the solution in interval notation.

16. $|ak-b|>b; k$
17. $|cp|\leq dp+c; p$
18. $|qm-r|-q\geq q-r; m$

19. Let $a > 0$, $b < 0$, and $x \neq 0$. Is $|ax| > |bx|$ a true inequality? Explain.

CHALLENGE

20. Let m, n, and p be positive integers such that $m < n$ and $n \geq p$. What is the relationship between $m - 1$ and p?

____

21. Let $b > 0$. What value(s) of x satisfy $|x + b| \geq |b - x|$?

____

SOLUTIONS

1. $(18, \infty)$
2. $(-\infty, 36]$
3. $(-\infty, 10]$
4. $(-\infty, 2)$
5. $(-\infty, -3]$
6. $(-\infty, \infty)$
7. $\left(\frac{6}{5}, \infty\right)$
8. $\left[\frac{b+c}{a}, \infty\right)$
9. 39,900 m^2
10. About 3 months
11. $(-20, 6)$
12. $(-\infty, -2] \cup [7, \infty)$
13. $(-\infty, \infty)$
14. $(-\infty, -4] \cup [4, \infty)$
15. $\varnothing$
16. $(-\infty, 0) \cup \left(\frac{2b}{a}, \infty\right)$
17. $\left[-\frac{c}{c+d}, \frac{c}{c-d}\right]$
18. $\left(-\infty, \frac{2r-2q}{q}\right] \cup [2, \infty)$
19. This is a false inequality. For example, let $a = 2$, $b = -2$ and $x = 1$. Then we have $|2(1)| > |-2(1)|$ or $2 > 2$, which shows the inequality is false. Other similar examples will suffice.
20. $m - 1 < p$ since $m < n$ and $n = p + 1$, where $n > p$.
21. $x \geq 0$

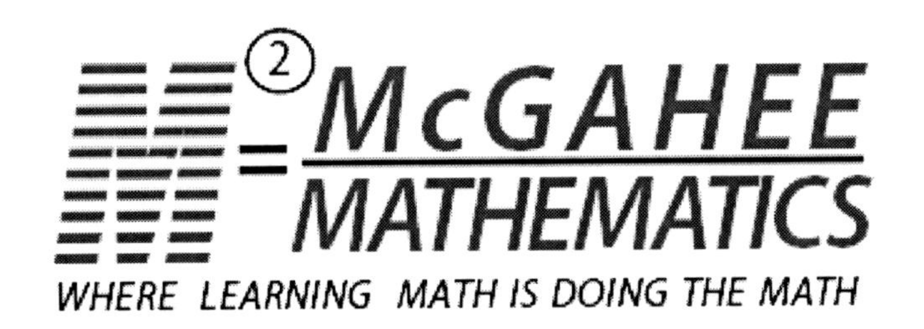

GRAPHING POINTS IN THE PLANE

QUESTIONS **ANSWERS**

Graph each point the Cartesian plane. State which quadrant or axis each point is located.

1. (2, 5) ___________
2. (-1, -3) ___________
3. (0, 4) ___________
4. (6, -10) ___________
5. (-7, 8) ___________
6. (9, 0) ___________
7. (0, -2) ___________
8. (-3, 0) ___________

SLOPE OF A LINE

Slope Formula: $m = \frac{y_2 - y_1}{x_2 - x_1}$

Find the slope of each line containing the following points.

9. (3, 4) and (-2, -1) ___________
10. (8, 6) and (-12, 6) ___________
11. (-5, 7) and (10, -2) ___________
12. (9, 1) and (9, 3) ___________

X-INTERCEPT AND Y-INTERCEPT OF A LINE

x-intercept: $(x, 0)$
y-intercept: $(0, y)$

Find the x-intercept and y-intercept of each line.

13. $2x + y = 4$ __________
14. $6x - 4y = 20$ __________
15. $3y - 15 = 0$ __________
16. $8x + 12y = -36$ __________

SLOPE-INTERCEPT FORM OF A LINE

Use the information in each question to write an equation of the line in slope-intercept form. Graph the line.

17. Slope of a line is $-\frac{1}{2}$ passing through (3, 0) __________
18. The line passing through (4, 7) and (3, 7) __________
19. Slope of a line is 1 passing through (5, -4). __________

CHALLENGE

20. Find an equation of a line in slope-intercept form that passes through the points $(1, y)$ and $(3, -2)$ with a slope of $4y - 1$.

____

21. Find the value of k such that the slope of the line $(2 - k)x - 3y = 0$ is equal to the slope of the line passing through $(0, k)$ and $(2, k - 4)$. What are the equations for both of these lines in slope-intercept form?

____

SOLUTIONS

Points for Questions 1-8 are plotted below.

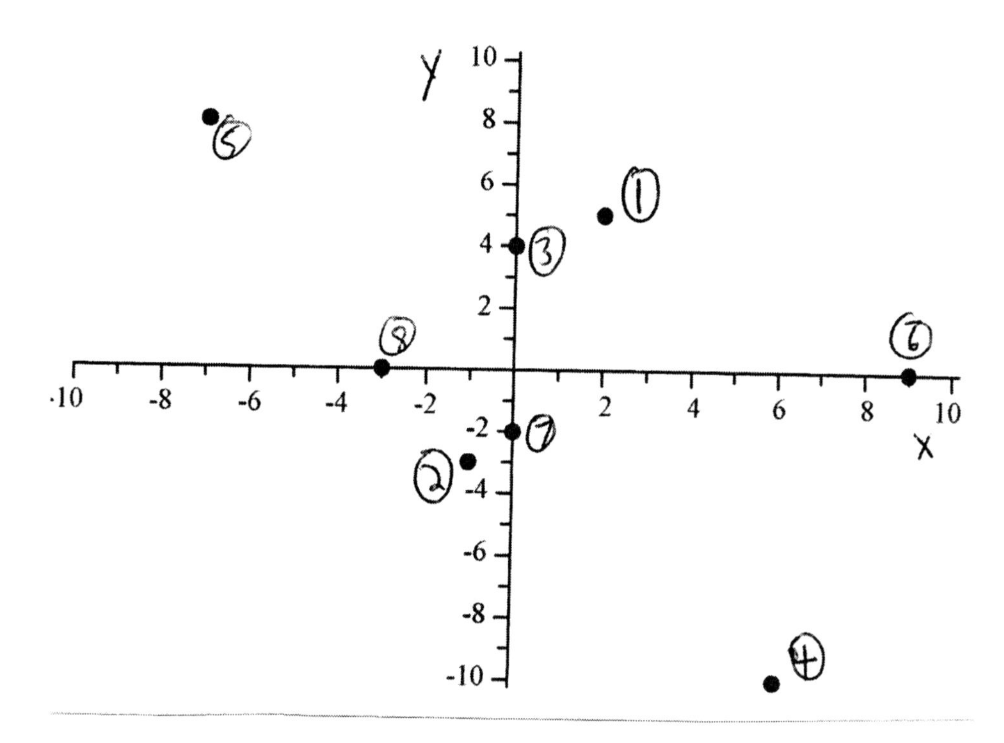

9. 1
10. 0
11. $-\frac{3}{5}$
12. Undefined
13. (2, 0) & (0, 4)
14. $\left(\frac{10}{3},0\right)$ & (0, -5)
15. No x-intercept & (0, 5)
16. $\left(-\frac{9}{2},0\right)$ & (0, -3)
17. $y=-\frac{1}{2}x+\frac{3}{2}$
18. $y=7$
19. $y=x-9$

Graphs for Questions 17-19 are below.

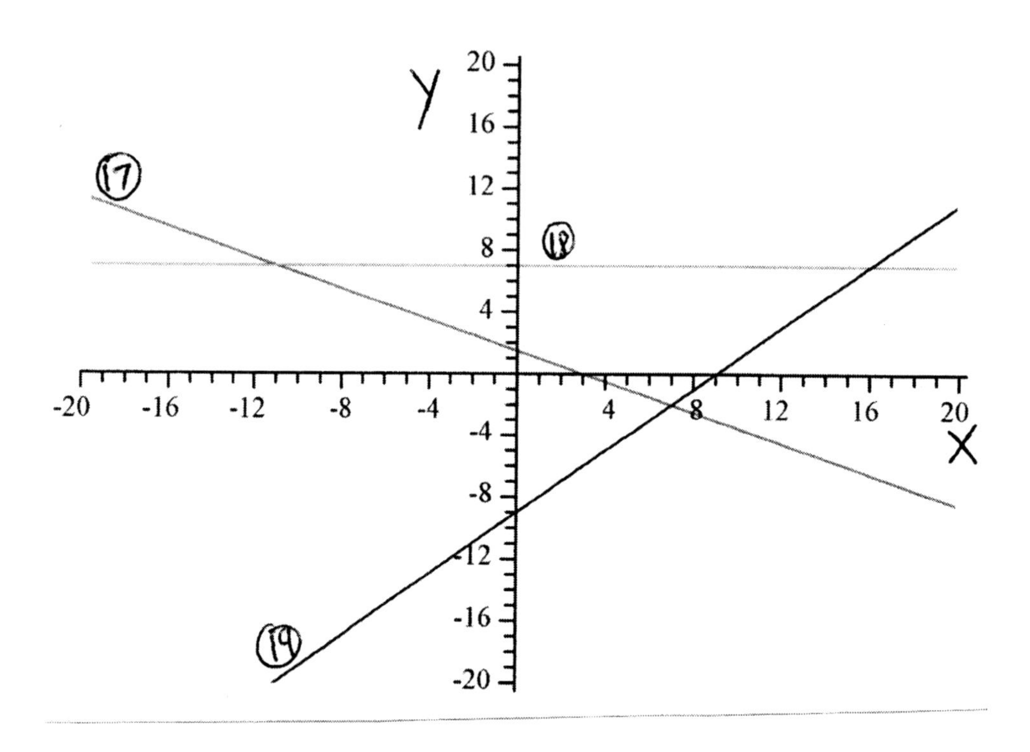

20. $y = -x + 1$

21. $k = 8$; $y = -2x$; $y = -2x + 8$

PARALLEL AND PERPENDICULAR LINES

QUESTIONS **ANSWERS**

Determine whether each pair of lines is parallel, perpendicular, or neither.

1. $y = 2x - 1, y = 2x + 3$ __________
2. $x + y = 0, y = x - 4$ __________
3. $3x - y = 6, 2y = -x + 5$ __________
4. $7x + 14y = 28, x + 2y = 10$ __________
5. $8y - 4x = 12, 12x + 8y = -48$ __________

Write an equation of a line for each given description.

6. Parallel to the line $x - 2y = 1$, passing through (3, 4)

____

7. Perpendicular to the line $5x + 10y = -15$, passing through (-1, 1)

____

8. Parallel to the line $3x + y = 9$, passing through the y-intercept of $3y - x = 12$

____

9. Perpendicular to $8y - 24x = 64$, passing through the x-intercept of $2x + y = -4$

____

SOLVING SYSTEMS OF LINEAR EQUATIONS

Solve each system by graphing.

10. $x - y = 1, y + x = 3$ ________________
11. $y = 2x$, $-2x + y = 4$ ________________
12. $3x - 6y = 9$, $-2y + x = 3$ ________________

Solve each system by substitution.

13. $x = y + 5, x + y = 7$ ________________
14. $2x - y = 4, y = -2x$ ________________
15. $y - 5x = 3$, $10x - 2y = -3$ ________________

Solve each system by elimination.

16. $-12x + 9y = 18$, $3x - 2y = -6$ ________________
17. $4x + 6y = 24$, $3y + 2x = 15$ ________________
18. $x - y = 1$, $-7y + 7x = 7$ ________________

Choose the best words from the word bank to fill in the blanks.

WORD BANK

Parallel, Intersect, Perpendicular, Coinciding, Horizontal, Slope

19. A system of linear equations has either only one solution if the lines _____________ one another at a single point, no solution if the lines are ____________, or infinitely many solutions for _______________ lines.

CHALLENGE

20. Write an equation of a line in standard form that is perpendicular to $y = mx + b$ passing through the point $(m , - b)$, where $m \neq 0$. Standard form is $Ax + By = C$, where A, B, and C are real numbers. If $A = 2$ and $C = -4$, then what point does this line pass through?

____

21. Solve the system of equations.

$$ax + by = c$$
$$dx + ey = f$$

$x =$ _____________________

$y =$ _____________________

SOLUTIONS

1. Parallel
2. Perpendicular
3. Neither
4. Parallel
5. Neither
6. $y = \frac{1}{2}x + \frac{5}{2}$
7. $y = 2x + 3$
8. $y = -3x + 4$
9. $y = -\frac{1}{3}x - \frac{2}{3}$
10. (2, 1)
11. No solution
12. Infinitely many solutions
13. (6, 1)
14. (1, -2)
15. No solution
16. (-6, -6)
17. No solution
18. Infinitely many solutions
19. Intersect, Parallel, Coinciding
20. $\frac{1}{m}x + y = 1 - b$, $\left(\frac{1}{2}, -5\right)$
21. $x = \frac{ce - bf}{ae - bd}$, $y = \frac{cd - af}{bd - ae}$

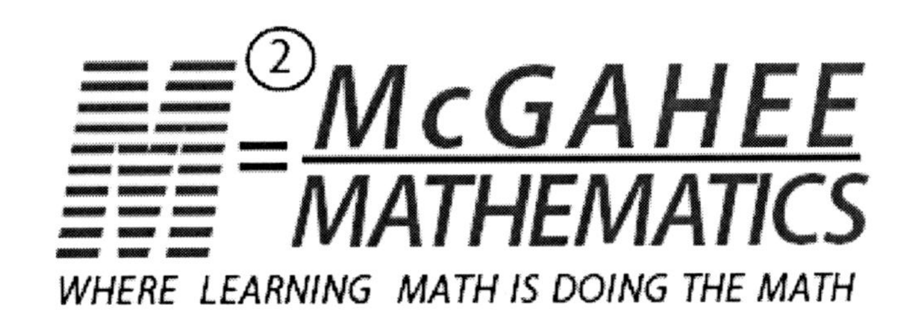

CHARACTERISTICS OF POLYNOMIALS

QUESTIONS **ANSWERS**

Find the degree and coefficient for each monomial.

1. $3x^2$ __________
2. $-8a$ __________
3. $12p^3q^5$ __________
4. 7 __________
5. $y^0z^6w^3$ __________

Write each polynomial in descending order. Find the degree and leading coefficient for each polynomial.

6. $8t^2-10-2t^3+4t$ __________
7. $m^5+3m-15m^4+25$ __________
8. $-5a^4+1-a^2$ __________
9. $36r^4-9r^7+54r^8-101+77r^3-r$ __________
10. $64-y^2$ __________

ADDING POLYNOMIALS

Add the polynomials. Write in descending order.

11. $(2x^4-5)+(9-x+6x^4)$ ____________________
12. $(13p^2-4p+1)+(-8p-7+5p^2)$ ____________________
13. $(z+2)+(2-2z)+(z-4)$ ____________________

14. $(38n^6 + 11n^4 - 24n + 19) + (4n - 40n^6 + 61 - 7n^4) + (-4n^4 + 2n^6)$

____

15. $(r^2 + 2rt) + (-4t^2 - rt - 3r^2) + (-3r^2 - rt + 4t^2)$

____

SUBTRACTING POLYNOMIALS

Subtract the polynomials. Write in descending order.

16. $(2x - 1) - (3x^2 + 10x - 5)$ ______________
17. $(-6z^3 - 12z + 8) - (-6z + 14z^2 - 20)$ ______________
18. $(7t + 35) - (35 - 40t + 2t^2)$ ______________
19. $(a^4 - 4a^3b + 8ab^2 - 16b^4) - (2a^3b + 15b^4 + 9a^4 - ab^2)$ ______________

CHALLENGE

20. A wooden stick is broken into 3 individual pieces. The second piece is 5 inches longer than the first piece. The third piece is 4 inches less than three times the square of the first piece. Write a polynomial that describes the sum of the three individual pieces. Let p represent the length of the first piece. What is the total length of the wooden stick if $p = 2$ inches?

____

21. Find the perimeter of the following geometric figure. Assume that the four-sided figure is a rectangle of width x and connected to a semi-circle with diameter d. If P is the perimeter of the figure, $d = y$, and $x = 2y$, then what is y in terms of P?

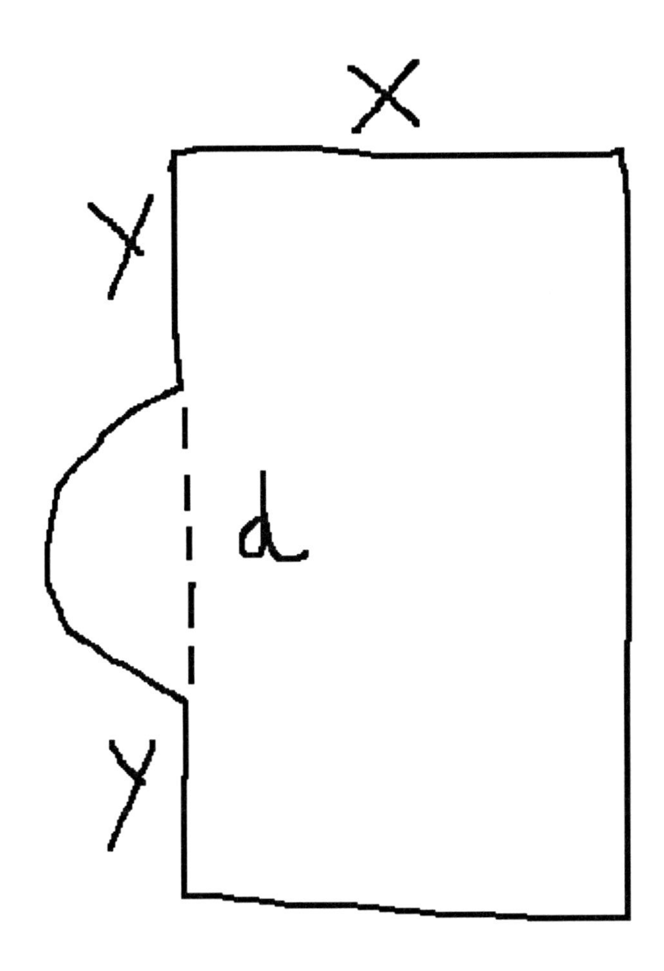

____

SOLUTIONS

1. Degree: 2, Coefficient: 3
2. Degree: 1, Coefficient: -8
3. Degree: 8, Coefficient: 12
4. Degree: 0, Coefficient: 7
5. Degree: 9, Coefficient: 1
6. $-2t^3 + 8t^2 + 4t - 10$, Degree: 3, Leading Coefficient: -2
7. $m^5 - 15m^4 + 3m + 25$, Degree: 5, Leading Coefficient: 1
8. $-5a^4 - a^2 + 1$, Degree: 4, Leading Coefficient: -5
9. $54r^8 - 9r^7 + 36r^4 + 77r^3 - r - 101$, Degree: 8, Leading Coefficient: 54
10. $-y^2 + 64$, Degree: 2, Leading Coefficient: -1
11. $8x^4 - x + 4$
12. $18p^2 - 12p - 6$
13. 0
14. $-20n + 80$
15. $-5r^2$
16. $-3x^2 - 8x + 4$
17. $-6z^3 - 14z^2 - 6z + 28$
18. $-2t^2 + 47t$
19. $-8a^4 - 6a^3b + 9ab^2 - 31b^2$
20. $3p^2 + 2p + 1$, 17 inches
21. $2x + 4y + \frac{\pi d}{2}$, $y = \frac{P}{\left(\frac{\pi}{2} + 9\right)}$

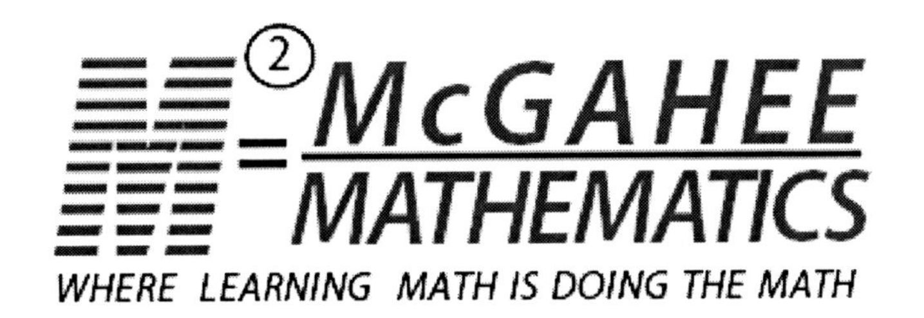

MULTIPLYING POLYNOMIALS

QUESTIONS **ANSWERS**

Multiply each of the following monomials and simplify.

1. $(2x^3)(5x)$ ____________
2. $(y^2)(-4y^5)$ ____________
3. $(-a)(3a^4b^2)(-6b^8)$ ____________
4. $(12m^7n^2p)(3np^4)(5mn^3)$ ____________

Distribute and simplify.

5. $7x(4x^2+8)$ ____________
6. $-6z^3(2z-5z^4)$ ____________
7. $3pq^2(9p^3-16p^2q+20q^3)$ ____________
8. $12rs^2t(-12rs^2+4r^3t+10r^4s^3t^5)$ ____________

Multiply each of the following binomials and simplify.

9. $(x+3)(x+2)$ ____________
10. $(y-1)(y+4)$ ____________
11. $(w+5)(w-5)$ ____________
12. $(b-7)(b-8)$ ____________
13. $(a+6)(a+6)$ ____________
14. $(2k+3)(k-5)$ ____________
15. $(4m-9)(11m+7)$ ____________

Distribute the binomial to the trinomial and simplify.

16. $(t+2)(t^2-4t+1)$ ____________

17. $(3x-4)(6x^3+12x^2-5)$ ____________

18. $(z+8)(z^2-2z-10)$ ____________

19. $(n-1)(n^2+n+1)$ ____________

CHALLENGE

20. A binomial squared is called a perfect square trinomial. If a and b are real numbers, then $(a+b)^2 = (a+b)(a+b) = a^2 + 2ab + b^2$ and $(a-b)^2 = (a-b)(a-b) = a^2 - 2ab + b^2$.

Here is the problem: Find the difference of the square of five more than 4 times a number and the square of five less than 4 times a number. Let x represent the number.

____

21. True or False: $(m+n)(m^2-mn+n^2)-(m-n)(m^2+mn+n^2)=2n^3$
Justify your answer.

____

SOLUTIONS

1. $10x^4$
2. $-4y^7$
3. $18a^5b^{10}$
4. $180m^8n^6p^5$
5. $28x^3+56x$
6. $-12z^4+30z^7$
7. $27p^4q^2-48p^3q^3+60pq^5$
8. $-144r^2s^4t+48r^4s^2t^2+120r^5s^5t^6$
9. x^2+5x+6
10. y^2+3y-4
11. w^2-25
12. $b^2-15b+56$
13. $a^2+12a+36$
14. $2k^2-7k-15$
15. $44m^2-71m-63$
16. t^3-2t^2-7t+2
17. $18x^4+12x^3-48x^2-15x+20$
18. $z^3+6z^2-26z-80$
19. n^3-1
20. $80x$
21. True.

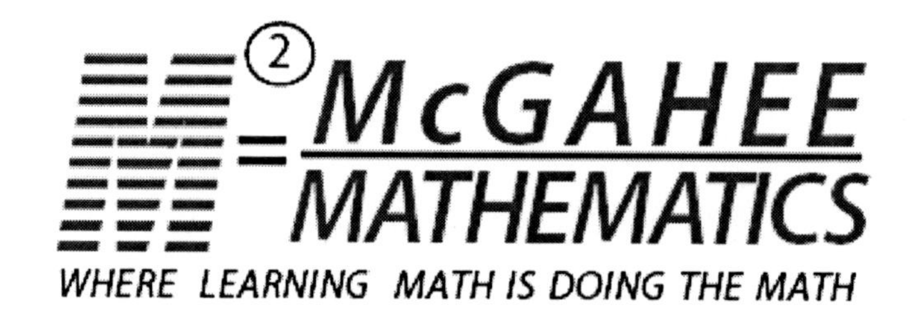

DIVIDING POLYNOMIALS

QUESTIONS **ANSWERS**

Divide each of the monomials.

1. $\dfrac{10x^5}{2x^2}$ __________

2. $\dfrac{8y^3}{14y}$ __________

3. $\dfrac{96ab^2}{3a^2b}$ __________

4. $\dfrac{125p^4q^2r^6}{20p^3qr^2}$ __________

5. $\dfrac{56m^{10}n^8}{14m^7n^3}$ __________

Divide each polynomial by the monomial.

6. $\dfrac{24x^3+36x^2}{4x}$ __________

7. $\dfrac{-5z^6+40z^4}{-10z^3}$ __________

8. $\dfrac{9y^2-12y+51}{3y^2}$ __________

9. $\dfrac{96r^2t-150rt^3-78r^4t^2}{6rt}$ __________

10. $\dfrac{a^4+4a^3b+6a^2b^2+4ab+b^4}{2a^2b^2}$ __________

Divide the polynomials.

11. $\dfrac{x^2+7x+12}{x+4}$ ____________________

12. $\dfrac{y^2-6y+8}{y-1}$ ____________________

13. $\dfrac{t^2-49}{t+7}$ ____________________

14. $\dfrac{21a^3+16a^2+7a-6}{3a-2}$ ____________________

15. $\dfrac{8p^3+125}{2p+5}$ ____________________

16. $\dfrac{14w^2+60w-72}{7w+9}$ ____________________

17. $\dfrac{2a^2+4ab+4b^2}{a+b}$ ____________________

18. $\dfrac{9x^2+24xy+16y^2}{3x+4y}$ ____________________

19. A rectangle has an area of a^2-64 square feet. If the length is $a+8$, then what is the perimeter of the rectangle?

____

CHALLENGE

20. Find the value of x such that $\dfrac{x^2-5x-66}{x-11}=\dfrac{3x+9}{12}$

____

21. The ratio of boys to girls in the classroom is 3 to 4. If there are 63 students in the classroom, how many are boys and how many are girls?

____

SOLUTIONS

1. $5x^3$
2. $\frac{4}{7y}$
3. $\frac{32b}{a}$
4. $\frac{25pqr^4}{4}$
5. $4m^3n^5$
6. $6x^2+9x$
7. $\frac{z^3}{2}+4z$
8. $3-\frac{4}{y}+\frac{17}{y^2}$
9. $16r-25t^2-13r^3t$
10. $\frac{a^2}{2b^2}+\frac{2a}{b}+3+\frac{2}{ab}$
11. $x+3$
12. $y-5+\frac{3}{y-1}$
13. $t-7$
14. $7a^2+10a+9+\frac{12}{3a-2}$
15. $4p^2-10p+25$
16. $2w+6-\frac{126}{7w+9}$
17. $2a+2b+\frac{2b^2}{a+b}$
18. $3x+4y$
19. $4a$
20. $x = -7$
21. 27 boys and 36 girls

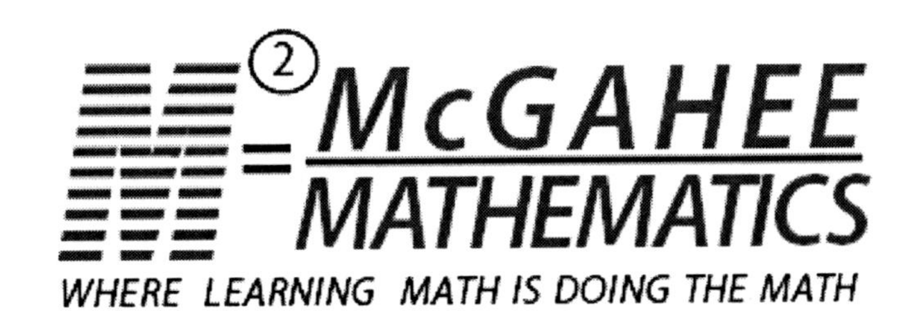

FACTORING POLYNOMIALS

QUESTIONS **ANSWERS**

Factor each polynomial by finding the greatest common factor.

1. $x^2 + 2x$ __________
2. $6y^3 - 8y^2$ __________
3. $15z^5 + 3z^4 + 27z$ __________
4. $20ab^2 - 16a^3b^4 + 32a^6b^5$ __________
5. $9mn + 45m^2n^2 - 54m^3n^3$ __________

Factor each trinomial. If the trinomial is prime, please state this.

6. $x^2 + 3x + 2$ __________
7. $y^2 - 5y - 14$ __________
8. $z^2 + 8z - 9$ __________
9. $a^2 - 10a + 21$ __________
10. $t^2 + 6t + 1$ __________
11. $8m^2 + 2m - 15$ __________
12. $12p^2 + 48p + 48$ __________
13. $7r^2 + 3r - 5$ __________

Factor each binomial or trinomial.

14. $4x^2 - 36$ __________
15. $y^3 + 8$ __________
16. $z^2 + 30z + 225$ __________
17. $t^4 - 1$ __________

18. $27n^6 - 64q^3$ ____________

19. $18w^2 - 84w + 98$ ____________

CHALLENGE

20. Find the sum of the smallest value of b and largest value of b such that $x^2 + bx - 20$ is factorable.

____

21. If $y^4 - r^4 = 8t^2$, $y + r = 2t$, and $y - r = t$, then what is $y^2 + r^2$?

____

SOLUTIONS

1. $x(x+2)$
2. $2y^2(3y-4)$
3. $3z(5z^4+z^3+9)$
4. $4ab^2(5-4a^2b^2+8a^5b^3)$
5. $9mn(1+5mn-6m^2n^2)$
6. $(x+1)(x+2)$
7. $(y-7)(y+2)$
8. $(z+9)(z-1)$
9. $(a-7)(a-3)$
10. *prime*
11. $(4m-5)(2m+3)$
12. $12(p+4)^2$
13. *prime*
14. $4(x+3)(x-3)$
15. $(y+2)(y^2-2y+4)$
16. $(z+15)^2$
17. $(t^2+1)(t+1)(t-1)$
18. $(3n^2-4q)(9n^4+12n^2q+16q^2)$
19. $2(3w-7)^2$
20. 0
21. 4

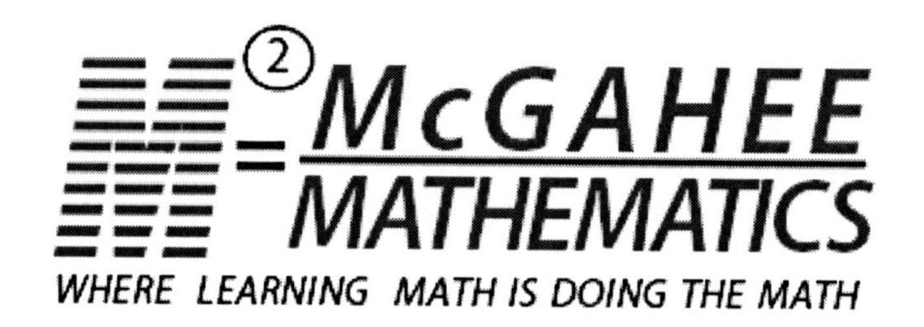

RATIONAL EXPRESSIONS

QUESTIONS **ANSWERS**

Find the value that makes the rational expression undefined.

1. $\dfrac{1}{x-3}$ ____________

2. $\dfrac{2y}{y+6}$ ____________

3. $\dfrac{4+z}{7-7z}$ ____________

4. $\dfrac{-5}{19a}$ ____________

Simplify each of the following rational expressions.

5. $\dfrac{6x-12}{8x-16}$ ____________

6. $\dfrac{y^2+4y}{y^2-16}$ ____________

7. $\dfrac{z^2+3z-40}{z^3-125}$ ____________

8. $\dfrac{t^4-1}{10t^2+20t+10}$ ____________

Multiply and Divide each of the following rational expressions.
Simplify completely.

9. $\frac{8x}{x+2} \bullet \frac{3x^3+6x^2}{24x^2}$ ____________

10. $\frac{12y^2-18y+6}{2y-1} \bullet \frac{y+1}{y^2-1}$ ____________

11. $\frac{z^3-27}{21-7z} \div \frac{5z^2+15z+45}{49z}$ ____________

12. $\frac{100-r^2}{r^4-81} \div \frac{-6r-60}{3-r}$ ____________

Add and Subtract each of the following rational expressions.
Simplify completely.

13. $\frac{2}{x-1} + \frac{2x-4}{x-1}$ ____________

14. $\frac{-3y+5}{y^2+5y+6} + \frac{4y-2}{y^2+5y+6}$ ____________

15. $\frac{z^2-7z}{z^2-64} - \frac{9z-64}{z^2-64}$ ____________

16. $\frac{2p^3}{p^2+10p+25} - \frac{p^3-125}{p^2+10p+25}$ ____________

17. $\frac{n+1}{n-2} + \frac{n-3}{n+4}$ ____________

18. $\frac{2m-6}{m^2+3m-18} - \frac{m+1}{m^2-9}$ ____________

Simplify the following rational expression.

19. $\frac{1}{q} + \frac{2}{q+1} - \frac{3}{q+2}$ ____________

CHALLENGE

20. Simplify the complex rational expression. What condition makes the rational expression equal to 0?

$$\frac{\frac{a}{b}-\frac{b}{a}}{a+b}$$ ____________________________

21. Simplify the complex rational expression.

$$\frac{x+y}{x^2-y^2} \div \frac{x-y}{x^3-y^3} - \frac{x^2+xy}{x-y} + \frac{y^2}{x+y}$$ ____________________________

SOLUTIONS

1. $x = 3$
2. $y = -6$
3. $z = 1$
4. $a = 0$
5. $\frac{3}{4}$
6. $\frac{y}{y-4}$
7. $\frac{z+8}{z^2-5z+25}$
8. $\frac{(t^2+1)(t-1)}{10(t+1)}$
9. x
10. 6
11. $\frac{-7z}{5}$
12. $\frac{10-r}{6(r^2+9)(r+3)}$
13. 2
14. $\frac{1}{y+2}$
15. $\frac{y-8}{y+8}$
16. $\frac{p^2-5p+25}{p+5}$
17. $\frac{2(n^2+5)}{(n-2)(n+4)}$
18. $\frac{m^2-7m-25}{(m+6)(m+3)(m-3)}$
19. $\frac{2(2q+1)}{q(q+1)(q+2)}$

20. $\frac{a-b}{ab}$ If $a = b$, then the rational expression is 0.

21. $\frac{2xy^2}{(x+y)(x-y)}$

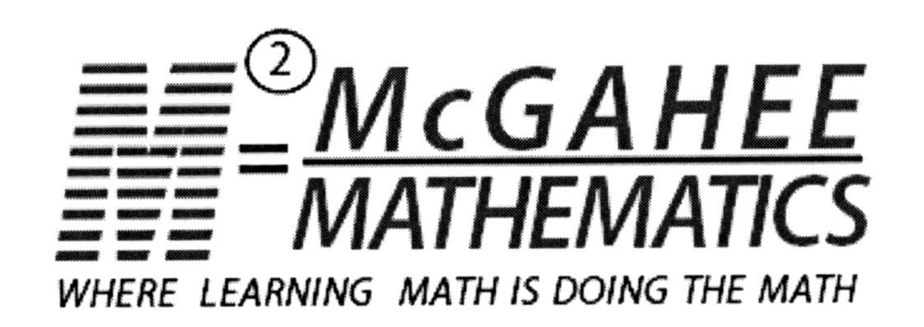

RATIONAL EQUATIONS

QUESTIONS **ANSWERS**

Solve each rational equation.

1. $\frac{2}{x}=\frac{3}{x+2}$ __________

2. $\frac{y}{6}=\frac{y-4}{2}$ __________

3. $\frac{1}{z}+\frac{5}{8}=\frac{15}{16}$ __________

4. $\frac{4}{a-1}+\frac{2}{a}=\frac{-2}{a^2-a}$ __________

5. $\frac{3t+9}{6}-\frac{1}{8}=\frac{t-3}{12}$ __________

6. $\frac{1}{r+2}-\frac{1}{r-2}=\frac{4}{4-r^2}$ __________

Solve each of the rational equations for the given variable.

7. $\frac{1}{R}=\frac{1}{R_1}+\frac{1}{R_2}$; R_1 __________

8. $\frac{1}{f}-\frac{1}{d}=\frac{1}{s}$; f __________

9. $\frac{a}{x}+\frac{b}{y}=\frac{c}{xy}$; y __________

10. $\frac{t-m}{n+m}-\frac{t+n}{m-n}=\frac{mt+n}{m^2-n^2}$; t __________

Solve each word problem.

11. The ratio of one more than a number and two is the ratio of four less than five times that number and six. What is the number?

____

12. A 5 foot woman stands 20 feet away from a 65 foot lamp that casts a shadow of an unknown length. What is the length of the shadow in feet and inches?

____

13. There are a certain number of red, blue, and green marbles inside a jar. If you remove three marbles from the jar, then the probability of choosing two green marbles is 1/3. How many marbles were originally in the jar?

____

14. A car travels 70 miles at some speed in the same time it can travel 90 miles at 10 miles per hour faster than that speed. What is the car's speeds and time to travel these distances?

____

15. It takes Bob 3 hours to paint the bedroom. Jenny can paint the bedroom in 5 hours. How long does it take for Bob and Jenny to paint the bedroom together in hours, minutes, and seconds?

____

Solve each conceptual problem.

16. If the rational equation $\frac{ax+b}{bx+a} = -1$ has the solution $x = 2$, then what is the value of $a + b$? ______________________________

17. Which value(s) of t cannot be a solution to this equation?

$$\frac{at}{t-b}+\frac{b}{t+a}=\frac{c}{t^2-(b-a)t-ab}$$

---

18. Solve the complex rational equation for x.

$$\frac{a}{1+\frac{x}{a+x}}=-a$$

---

19. Solve the rational equation for R.

$$\frac{1}{R}=R_1+R_2+\frac{1}{r}$$

CHALLENGE

20. If $\frac{1}{x}+\frac{1}{y}+\frac{1}{z}=xy$, then show that $z=\frac{1}{xy-\left(\frac{1}{x}+\frac{1}{y}\right)}$.

---

21. If $\frac{s^2}{a^2}-\frac{t^2}{b^2}=1$, and $\frac{s}{a}+\frac{t}{b}=-1$, then show that $bs=a(t-b)$.

SOLUTIONS

1. $x = 4$
2. $y = 6$
3. $z = \frac{16}{5}$
4. No solution
5. $t = \frac{-39}{10}$
6. r is any real number
7. $R_1 = \frac{RR_2}{R_2 - R}$
8. $f = \frac{ds}{d + s}$
9. $y = \frac{c - bx}{a}$
10. $t = \frac{m^2 + 2mn + n^2 - n}{m + 2n}$
11. $\frac{7}{2}$
12. 1 feet, 8 inches
13. 9 marbles
14. 35 mph for 70 miles, 45 mph for 90 miles, 2 hours
15. 1 hour, 52 minutes, 30 seconds.
16. 0
17. $t = b$, $t = -a$
18. $x = -\frac{2a}{3}$
19. $R = \frac{r}{r(R_1 + R_2) + 1}$
20. Hint: Solve for z and divide both the numerator and denominator by the right hand side of the original equation.
21. Hint: Factor the left hand side of the original equation.

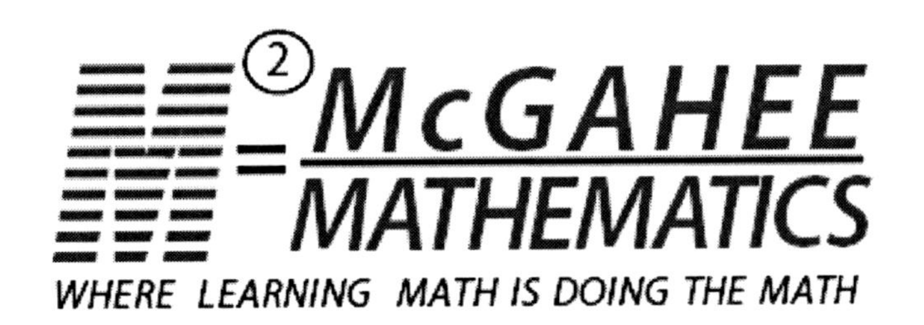

RADICAL EXPRESSIONS WITHOUT VARIABLES

QUESTIONS **ANSWERS**

Simplify each of the following radical expressions.

1. $\sqrt{4}$ ________
2. $-\sqrt{25}$ ________
3. $\sqrt{80}$ ________
4. $\sqrt[3]{-1}$ ________
5. $\sqrt[3]{54}$ ________

Combine like radicals and simplify.

6. $6\sqrt{8}+9\sqrt{2}$ ________
7. $-4\sqrt{3}+10\sqrt{5}+7\sqrt{48}$ ________
8. $\sqrt[3]{162}+3\sqrt[3]{384}$ ________
9. $2\sqrt[3]{7}-6\sqrt[3]{56}+21\sqrt[3]{135}$ ________

Simplify each radical expression.

10. $\sqrt{2}\bullet\sqrt{6}$ ________
11. $\sqrt{45}\bullet\sqrt{10}$ ________
12. $\sqrt[3]{256}\bullet\sqrt[3]{80}$ ________
13. $\sqrt[3]{-3,000}\bullet\sqrt[3]{630}$ ________
14. $\sqrt{\frac{49}{121}}$ ________
15. $\sqrt[3]{\frac{512}{1728}}$ ________

Write each radical expression in terms of a rational exponent. Simplify completely.

16. $\sqrt{3}$ __________

17. $\sqrt{125}$ __________

18. $\sqrt[3]{256}$ __________

19. $\sqrt[3]{2187}$ __________

CHALLENGE

20. The number $\frac{\sqrt{3}+\sqrt{5}}{2}$ exists between what two rational numbers?

____

21. $\sqrt{8}$ is how many times larger than $\sqrt[3]{16}$?

____

SOLUTIONS

1. 2
2. -5
3. $4\sqrt{5}$
4. -1
5. $3\sqrt[3]{2}$
6. $21\sqrt{2}$
7. $24\sqrt{3}+10\sqrt{5}$
8. $15\sqrt[3]{6}$
9. $53\sqrt[3]{7}$
10. $2\sqrt{3}$
11. $15\sqrt{2}$
12. $16\sqrt[3]{5}$
13. $-30\sqrt[3]{70}$
14. $\frac{7}{11}$
15. $\frac{2}{3}$
16. $3^{\frac{1}{2}}$
17. $5^{\frac{3}{2}}$
18. $4^{\frac{4}{3}}$
19. $3^{\frac{7}{3}}$
20. $\frac{3}{2}$ & $\frac{5}{2}$
21. $\sqrt[6]{2}$

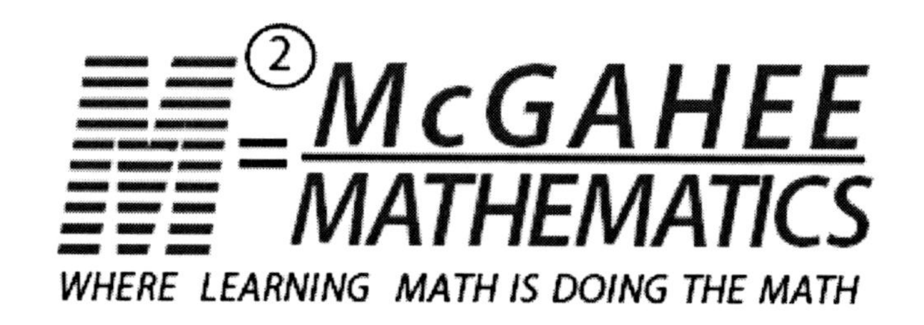

RADICAL EXPRESSIONS WITH VARIABLES

QUESTIONS **ANSWERS**

Simplify each radical expression. Assume that the variables can be any real number.

1. $\sqrt{9x^3}$ ________
2. $\sqrt{36y^8}$ ________
3. $\sqrt[3]{54z^5}$ ________
4. $\sqrt[3]{432t^{10}}$ ________

Expand each radical expression. Simplify completely.

5. $(2+\sqrt{x})(4+3\sqrt{x})$ ________
6. $(\sqrt{y}-1)(\sqrt{y}+1)$ ________
7. $(3+5\sqrt{8z})^2$ ________
8. $(\sqrt{s}-\sqrt{t})(s+\sqrt{st}+t)$ ________

Rationalize the denominator. Simplify completely.

9. $\dfrac{1}{\sqrt{x}}$ ________
10. $\dfrac{2y}{\sqrt{y^3}}$ ________
11. $\dfrac{5}{\sqrt{z}+3}$ ________
12. $\dfrac{\sqrt{6p}+\sqrt{10q}}{\sqrt{6p}-\sqrt{10q}}$ ________

Rationalize the numerator. Simplify completely.

13. $\dfrac{\sqrt{4x}}{8x}$ __________

14. $\dfrac{\sqrt{2y}-1}{y+2}$ __________

15. $\dfrac{\sqrt{3z^5}}{9z-7}$ __________

16. $\dfrac{\sqrt{5t}+\sqrt{30w}}{\sqrt{30t}+\sqrt{5w}}$ __________

Simplify each of the abstract radical expressions.

17. $\dfrac{a}{x+\sqrt{y}}+\dfrac{a}{x-\sqrt{y}}$ __________

18. $\dfrac{b}{\sqrt{t}+\sqrt{r}}-\dfrac{a+b}{\sqrt{t}-\sqrt{r}}$ __________

19. $\dfrac{a\sqrt{n}+b\sqrt{m}}{a}+\dfrac{a\sqrt{n}+b\sqrt{m}}{b}$ __________

CHALLENGE

20. Factor $x^2-xy+\sqrt{x}+\sqrt{y}$ using difference of two squares with radicals.

____

21. True or False.

$$\dfrac{\dfrac{\sqrt{a+b}}{a}+\dfrac{b}{\sqrt{a+b}}}{\dfrac{b}{\sqrt{a+b}}-\dfrac{\sqrt{a+b}}{a}}=\dfrac{b(a+1)+a}{b(a-1)-a}$$

____

SOLUTIONS

1. $3|x|\sqrt{x}$

2. $6y^4$

3. $3z\sqrt[3]{2z^2}$

4. $6t^3\sqrt[3]{2t}$

5. $8+10\sqrt{x}+3x$

6. $y-1$

7. $9+60\sqrt{2z}+200z$

8. $5\sqrt{5}-t\sqrt{t}$

9. $\dfrac{\sqrt{x}}{x}$

10. $\dfrac{2\sqrt{y}}{y}$

11. $\dfrac{5\sqrt{z}-15}{z-9}$

12. $\dfrac{3p+2\sqrt{5q}+5q}{3p-5q}$

13. $\dfrac{x}{4\sqrt{x}}$

14. $\dfrac{2y-1}{(y+2)\sqrt{2y}+y+2}$

15. $\dfrac{3z^3}{(9z-7)\sqrt{3z}}$

16. $\dfrac{t-6w}{(t-w)\sqrt{6}-5\sqrt{tw}}$

17. $\frac{2ax}{x^2 - y}$

18. $\frac{(a+2b)\sqrt{r} + a\sqrt{t}}{r-t}$

19. $\frac{(a+b)(b\sqrt{m} + a\sqrt{n})}{ab}$

20. $(\sqrt{x} + \sqrt{y})[x(\sqrt{x} - \sqrt{y}) + 1]$

21. True.

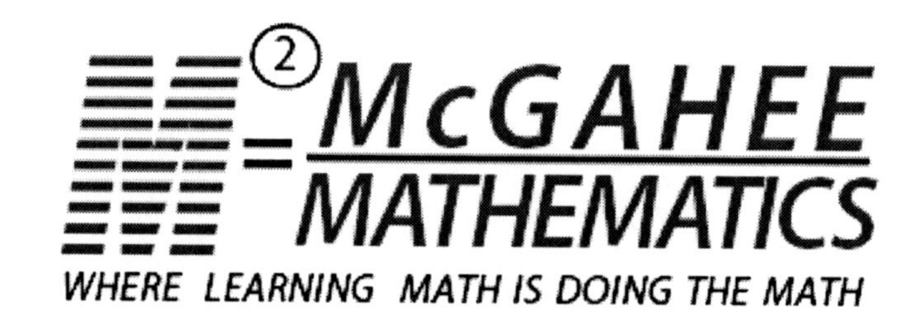

COMPLEX NUMBERS

QUESTIONS **ANSWERS**

Simplify by adding and subtracting complex numbers.

Write in $a + bi$ form.

1. $5+4i-8-3i$ __________
2. $12-7i-6+21i$ __________
3. $(2-i)^2+10+9i$ __________

Multiply complex numbers and simplify. Write in $a + bi$ form.

4. $(7+8i)(6+5i)$ __________
5. $(4+3i)(4-3i)$ __________
6. $(1-i)(2+i)$ __________
7. $(-4-7i)(6-11i)$ __________

Divide complex numbers and simplify. Write in $a + bi$ form.

8. $\frac{2+5i}{3+4i}$ __________
9. $\frac{1-i}{1+i}$ __________
10. $\frac{-9+3i}{6i}$ __________
11. $\frac{4}{8-5i}$ __________

Simplify and write each complex number in $a + bi$ form.

12. $\sqrt{-25}$ ____________

13. $\sqrt{-4} + \sqrt{-8}i$ ____________

14. $2\sqrt{-9} - 6i + 3i\sqrt{-100}$ ____________

Use the powers of i to simplify each complex number.

15. $-i^8$ ____________

16. $5i^{39}$ ____________

17. $24i^{16}$ ____________

Simplify each abstract complex number. Write in $a + bi$ form.

18. $\frac{a}{a+bi} + \frac{a}{a-bi}$ ____________

19. $\frac{a+bi}{ai} + \frac{a-bi}{ai}$ ____________

CHALLENGE

20. Show that $\frac{ai}{a-bi}$ can be written in the form $c(b - ai)$, where c is a real number. What is the value of c?

____

21. Let a, b, and c be positive real numbers. If $a^2 + b^2 = c$, then find the possible values of a and b in terms of c that satisfy this equation.

____

SOLUTIONS

1. $-3+i$
2. $6+14i$
3. $13+5i$
4. $2+83i$
5. 25
6. $3-i$
7. $-101+2i$
8. $\frac{26}{25}+\frac{7}{25}i$
9. $-i$
10. $\frac{1}{2}+\frac{3}{2}i$
11. $\frac{32}{89}+\frac{20}{89}i$
12. $5i$
13. $-2\sqrt{2}+2i$
14. -30
15. -1
16. $-5i$
17. 24
18. $\frac{2a^2}{a^2+b^2}$
19. $-2i$
20. $c=-\frac{a}{a^2+b^2}$
21. $a=\frac{c+1}{2}$, $b=\left(\frac{c-1}{2}\right)i$, $b=\left(\frac{1-c}{2}\right)i$

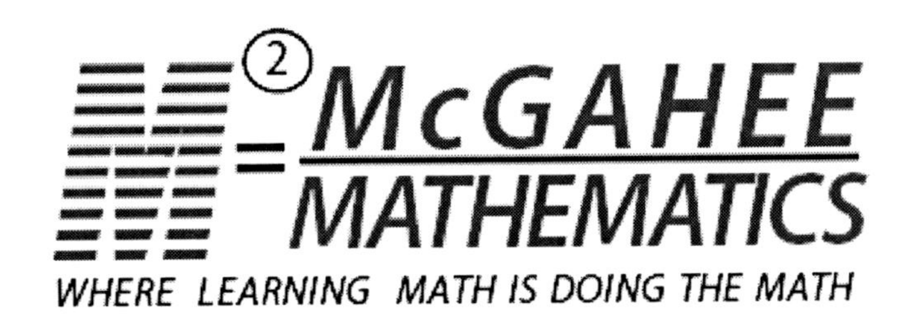

QUADRATIC EQUATIONS

QUESTIONS **ANSWERS**

Solve each quadratic equation by factoring.

1. $x^2 - 2x = 0$ __________
2. $3y^2 + 9y = 0$ __________
3. $z^2 + 4z - 5 = 0$ __________
4. $8a^2 = 16$ __________
5. $2t^2 + 17t + 21 = 0$ __________
6. $p^2 = -2p - 1$ __________
7. $-4r^2 + 8r + 12 = 0$ __________

Solve each quadratic equation by completing the square.

8. $x^2 + 6x = 11$ __________
9. $y^2 - 10y = 22$ __________
10. $7z^2 + 2z + 1 = 0$ __________
11. $8t^2 - 24t + 6 = 0$ __________

Solve each quadratic equation by the quadratic formula.

12. $2x(x - 5) = 3x + 1$ __________
13. $y(4 - y) + 5y = -6y + 8$ __________
14. $z^2 = -z(z - 1) - 2$ __________

Solve each abstract quadratic equation.

15. $A = 2\pi r^2 + 2\pi rh$; r ______________

16. $d = \dfrac{v^2 - v_o^{\,2}}{2a}$; v Assume that $v > 0$. ______________

Solve each of the following word problems.

17. The reciprocal of 1 more than a number is the ratio of that number and 2. Find the number(s).

____

18. Brett canoes 8 miles upstream and 8 miles downstream for a total time of 3 hours. If the speed of the current is 2 miles per hour, then what is the speed of the canoe in still water?

____

19. The perimeter of a rectangular garden is 90 feet. The area of the garden is 450 square feet. What are the dimensions of the garden?

____

CHALLENGE

20. A poster board of 8 inches x 10 inches needs an unknown margin length from all of the sides to create a frame for a picture. If the area of the picture is 1/10 the area of the poster board, then what are the dimensions of the picture?

____

21. The equation in physics $x = x_o + v_o t + \frac{1}{2}at^2$ describes the position x of an object in terms of its initial position x_o, initial velocity v_o, acceleration a, and time t.

Suppose that a boy who is 60 meters behind a school bus traveling at an initial velocity 20 meters per second accelerates to 4 meters per second squared. At what time will the boy catch up to the school bus? Round to the nearest second.

____

SOLUTIONS

1. $x = 0$ or $x = 2$
2. $y = 0$ or $y = -3$
3. $z = -5$ or $z = 1$
4. $a = \sqrt{2}$ or $a = -\sqrt{2}$
5. $t = -\frac{3}{2}$ or $t = -7$
6. $p = -1$
7. $r = -1$, or $r = 3$
8. $x = -3 \pm 2\sqrt{5}$
9. $y = 5 \pm \sqrt{47}$
10. $z = \frac{-1 \pm \sqrt{6}i}{7}$
11. $t = \frac{3 \pm \sqrt{6}}{2}$
12. $x = \frac{13 \pm \sqrt{177}}{4}$
13. $y = \frac{15 \pm \sqrt{193}}{2}$
14. $z = \frac{1 \pm \sqrt{15}i}{4}$
15. $r = \frac{-\pi h \pm \sqrt{\pi^2 h^2 + 2\pi A}}{2\pi}$
16. $v = \sqrt{v_o^{\,2} + 2ad}$
17. $x = -2$ or $x = 1$
18. 6 mph
19. 15 feet x 30 feet
20. 2 inches x 4 inches
21. 12 seconds

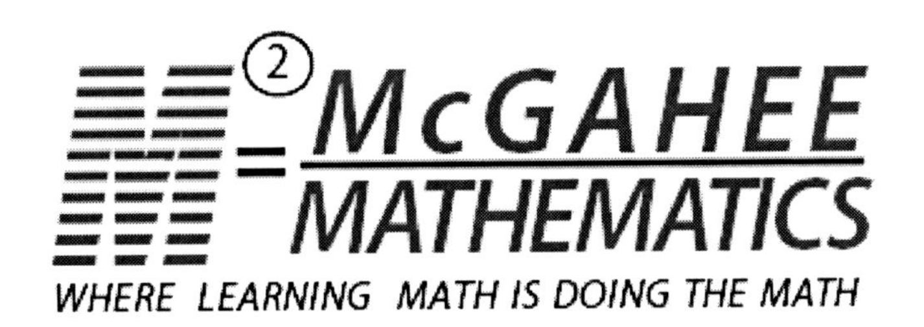

QUADRATIC INEQUALITIES

QUESTIONS **ANSWERS**

Solve each of the following quadratic inequalities.

1. $x^2 + 3x - 28 \geq 0$ ____________
2. $y^2 - 6y < 0$ ____________
3. $z^2 + 4z + 4 \leq 0$ ____________
4. $a^2 - 10a - 39 \leq 0$ ____________
5. $n^2 \geq 2 - n$ ____________
6. $p^2 \leq 9$ ____________
7. $2t^2 + 7t + 3 > 0$ ____________
8. $5m^2 < 125$ ____________
9. $13q - 6 \leq 8 - 12q^2$ ____________
10. $121r^2 - 22r + 1 \geq 0$ ____________

Solve each of the following abstract quadratic inequalities for the given variables. Assume that the letters in alphabetical order are numbered from least to greatest.

11. $x^2 - ax \leq bx - ab$; x ____________
12. $y^2 + cy > d(c + y)$; y ____________
13. $z^2 - 2rz < -r^2$; z ____________
14. $at^2 + bt \geq 0$; t ____________
15. $n^2 \leq \frac{p^2}{q^2}$; n ____________
16. $w(w + i) > -iw$; w ____________

Solve each word problem.

17. The sum of twice a number and its square is greater than eight. Find the numbers.

____

18. The height of a ball falling off a building is given by the following equation $h = -16t^2 + 64$, where the height h is in feet and the time t is in seconds. During what time interval will the ball be lower than 48 feet?

____

19. The perimeter of a rectangle is 40 meters. Find the range of dimensions of the rectangle if its area is at least 96 square meters.

____

CHALLENGE

20. Solve the inequality for v. $\frac{v^2}{c} - v \leq cv^2 + v$ Assume $0 < c < 1$.

____

21. Solve the inequality for x. $ax^3 - x^2 \geq x^2 - ax$ Assume $0 < a < 1$.

____

SOLUTIONS

1. $(-\infty,-7]\cup[4,\infty)$
2. (0, 6)
3. -2
4. [-3, 13]
5. $(-\infty,-2]\cup[1,\infty)$
6. [-3, 3]
7. $(-\infty,-3)\cup\left(-\frac{1}{2},\infty\right)$
8. (-5, 5)
9. $\left[-\frac{7}{4},\frac{2}{3}\right]$
10. $(-\infty,\infty)$
11. $[a, b]$
12. $(-\infty,-c)\cup(d,\infty)$
13. No solution
14. $\left(-\infty,-\frac{b}{a}\right)\cup[0,\infty)$
15. $\left[-\frac{p}{q},\frac{p}{q}\right]$
16. $(-\infty,-2i)\cup(0,\infty)$
17. $(-\infty,-4)\cup(2,\infty)$
18. $(1,2]$ seconds
19. The length and width are between 8 meters and 12 meters inclusively.
20. $\left[0,\frac{2c}{1-c^2}\right]$, $(-\infty,0]\cup\left[\frac{1-\sqrt{1-a^2}}{a},\infty\right)$

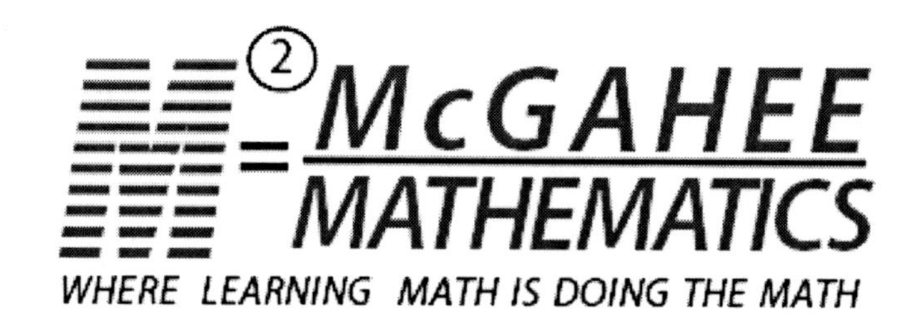

RADICAL EQUATIONS

QUESTIONS **ANSWERS**

Solve each of the following radical equations.

1. $\sqrt{x} = 2$ __________
2. $3\sqrt{y} - 4 = 8$ __________
3. $\sqrt{6z - 3} = -1$ __________
4. $\sqrt{2t + 1} + 4 = -5$ __________
5. $1 + \sqrt{a} = \sqrt{a + 7}$ __________
6. $\sqrt{m + 3} - 2\sqrt{m} = \sqrt{m - 1}$ __________
7. $\sqrt{2 - p} = p + 10$ __________
8. $\sqrt{r^2 + 9} = \sqrt{25 - r^2}$ __________
9. $\sqrt{4w} = w + \sqrt{w}$ __________
10. $\sqrt[3]{x} = -3$ __________
11. $\sqrt[3]{5y} + 25 = 30$ __________
12. $\sqrt[3]{z - 4} = -\sqrt[3]{7z + 2}$ __________
13. $\sqrt[3]{s^2 - 1} - \sqrt[3]{s + 1} = 0$ __________
14. $2\sqrt[3]{t^2 + 2} = 14$ __________

Solve each abstract radical equation for the given variable.

15. $\sqrt{x + a} = \sqrt{x} + \sqrt{a}$; x __________
16. $\sqrt{y - b} = \sqrt{y} - \sqrt{b}$; y __________
17. $\sqrt{z} + \sqrt{z + c} = \sqrt{z - c}$; z __________

Solve each word problem.

18. The speed v of a falling object is directly proportional to the square root of the distance d traveled, i.e. $v = k\sqrt{d}$, where k is a constant of proportionality. A ball falling off a 100 foot building falls at a speed of 80 feet per second. How far does a ball travel downward at a speed of 40 feet per second?

____

19. The power P (watts) in an electrical circuit is the product of the square of the current I (amps) and resistance R (ohms). If the power of an electrical circuit is 500 watts with a total resistance of 75 ohms, then what is the estimated current in the circuit? Round to the nearest whole number.

____

CHALLENGE

20. Solve the radical equation for x.

$$\sqrt{x} - a\sqrt[4]{x} = a$$

____

21. Solve the radical equation for y.

$$\sqrt{y + \sqrt{y - b}} = \sqrt{y + b - \sqrt{y + b}}$$

____

SOLUTIONS

1. $x = 4$
2. $y = 16$
3. No solution
4. No solution
5. $a = 9$
6. $m = 1$
7. $p = -7$
8. $r = \pm 2\sqrt{2}$
9. $w = 1$
10. $x = -27$
11. $y = 25$
12. $z = \frac{1}{4}$
13. $s = -1$ or $s = 2$
14. $t = \pm\sqrt{341}$
15. $x = 0$
16. $y = b$
17. $z = \pm\frac{2c\sqrt{3}}{3}$
18. 25 feet
19. 3 amps
20. $x = \left(\frac{a \pm \sqrt{a^2 + 4a}}{2a}\right)^4$
21. $y = \frac{b^2 + 4}{4}$

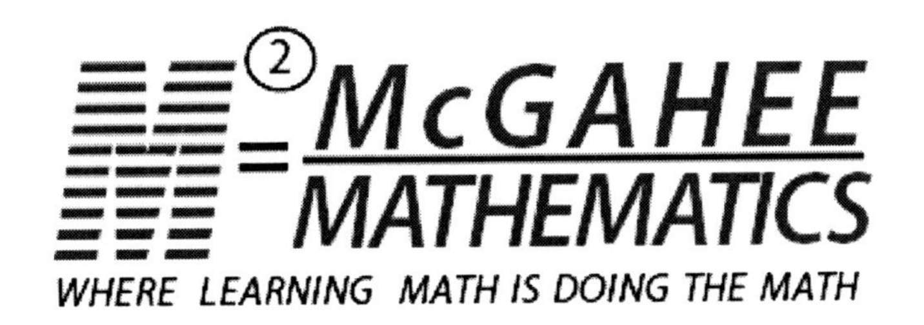

RATIONAL INEQUALITIES

QUESTIONS **ANSWERS**

Solve each of the following rational inequalities. State the solution in interval notation.

1. $\frac{1}{x-2} > 0$ __________

2. $\frac{1}{y} + \frac{1}{y+1} \le 0$ __________

3. $\frac{z^2 - 9}{z^2 + 3z - 40} \ge 0$ __________

4. $\frac{2t+5}{t+6} < \frac{1}{t}$ __________

5. $\frac{1}{r^2} > \frac{2}{r}$ __________

6. $\frac{4}{w-4} \ge \frac{8}{w+4}$ __________

7. $\frac{|a+1|}{a-1} < 0$ __________

8. $\frac{m^3 + 5m^2 + 6m}{m^2 - 10m} \ge 0$ __________

Solve each abstract rational inequality for the following variable. Assume all constants are positive.

9. $\frac{x-a}{x+a} \le 0; x$ __________

10. $\frac{y^2 - b^2}{y-b} > 0; y$ __________

11. $\frac{|cz+d|}{c-dz} \geq 0$; z, where $c < d$ ____________

12. $\frac{t}{k} < \frac{k}{t}$; t ____________

13. $\frac{ev}{v+e} + v > 0$; v ____________

14. $\frac{fw}{w-f} \leq w$; w ____________

Determine whether each statement is true or false. Circle the correct answer. If the statement is false, find a counterexample to show why it is false.

15. If $\frac{1}{x} > 1$, then $0 < x < 1$. True OR False

16. If $\frac{1}{y} < \frac{1}{|y|}$, then $y < 0$. True OR False

17. For all $z > 0$, $\frac{1}{z^2} > \frac{1}{z^3}$. True OR False

18. $-\frac{|z+9|}{|9-z|} \geq 0$ has no solution. True OR False

19. $\frac{1}{t} - t \leq 0$ has the solution $[-1,0) \cup [1,\infty)$. True OR False

CHALLENGE

20. Solve $\frac{x^2+(a+b)x+ab}{x-c} \geq 0$, where $0 < a < b < c$.

____

21. Solve $\frac{\sqrt{cx+d}}{cx-d} \leq 0$, where $0 < c < d$.

____

SOLUTIONS

1. $(2,\infty)$
2. $(-\infty,-1)\cup\left[-\frac{1}{2},0\right)$
3. $(-\infty,-8)\cup[-3,3]\cup(5,\infty)$
4. $(-6,-3)\cup(0,1)$
5. $(-\infty,0)\cup\left(0,\frac{1}{2}\right)$
6. $(-\infty,-4)\cup(4,12]$
7. $(-\infty,-1)\cup(-1,1)$
8. $[-3,-2]\cup(10,\infty)$
9. $(-a,a]$
10. $(-b,b)\cup(b,\infty)$
11. $\left(-\infty,\frac{c}{d}\right)$
12. $(-\infty,-k)\cup(0,k)$
13. $(-2e,-e)\cup(-e,0)\cup(0,e)$
14. $[0,f)\cup[2f,\infty)$
15. True.
16. True.
17. False. For example let $z=\frac{1}{2}$. This implies that $4 > 8$, which shows the rational inequality is false. Other examples will suffice.
18. False. For example, $z = -9$ is a solution to the rational inequality.
19. True.
20. $[-b,-a]\cup(c,\infty)$

    $\left[-\frac{d}{c},\frac{d}{c}\right)$

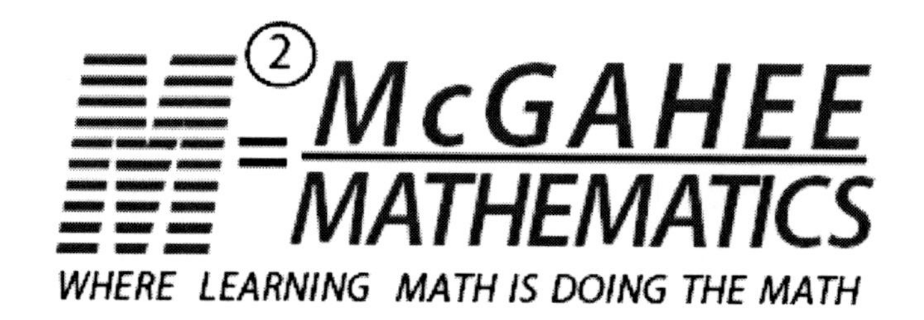

MIDPOINT/DISTANCE FORMULAS & PYTHAGOREAN THEOREM

QUESTIONS **ANSWERS**

Find the midpoint between the two given points.

1. (1, 1) and (3, 5) ________
2. (-2, -4) and (4, -6) ________
3. (0, 7) and (8, 0) ________
4. (10, -9) and (12, -9) ________
5. (-35, -40) and (-28, -17) ________

6. Find the values of x and y such that the midpoint of (x, -2) and (4, y) is (8, -10).

Find the distance between the two given points.

7. (-1, 2) and (2, -3) ________
8. (4, 0) and (0, 3) ________
9. (0, 0) and (-8, 6) ________
10. (-5, -7) and (-9, -10) ________
11. (-2, 4) and (-4, 2) ________

12. Find the largest value of a such that the distance between $(3, a)$ and $(-2, 1)$ is $5\sqrt{2}$.

___

Suppose a and b are legs and c the hypotenuse of a right triangle. Find the missing side.

13. $a = 3, b = 4$ $c =$ ________
14. $a = 5, c = 10$ $b =$ ________
15. $b = 7, c = 7\sqrt{2}$ $a =$ ________

16. The sides of a right triangle are $x + 2$ and $x + 4$ and the hypotenuse is $x + 6$. Find the perimeter and area of the right triangle. What do you notice about these values?

___

17. John and Carrie begin their run at the same point at 12:30 p.m., but in different directions. John runs east at a speed of 5 mph while Carrie runs north at a speed of 4 mph. About what time will they be 20 miles apart? Round to the nearest hour and minute.

___

18. The vertices of a right triangle XYZ are $X(0,0)$, $Y(0, b)$ and $Z(c, 0)$.
    a) Find the midpoint M of line segment YZ. ____________
    b) Show that M is equidistant from the vertices of right triangle XYZ.

___

19. Are the points $P(4, 0)$, $Q(1, 9)$, and $R(-3, 21)$ collinear? If the points P, Q, and R are collinear, show that $d(P, Q) + d(Q, R) = d(P, R)$.

___

CHALLENGE

20. Let the y-axis serve as a mirror that reflects each point to the right of the axis into a point to the left of the axis.

    Consider the points $A(2, 2)$ and $B(5, 1)$.

    a) Find the reflected points A' and B'. ________________
    b) Create the segments AA', AB, BB' and $A'B'$. What kind of figure did you create? ________________________
    c) What is the area of this figure? ________________________

21. Consider the points $M(a, a)$, $N(a, 3)$ and $O(2a, a)$, where $0 < a \leq 1$ that forms the triangle MNO.
    a) What type of triangle is this? ______________________
    b) If the value of a is increased by 1, find the coordinates of triangle $M'N'O'$ that yield a maximum area. ________________
    c) Find the ratio of the area of triangle MNO to triangle $M'N'O'$ in terms of a. ______________________

SOLUTIONS

1. (2, 3)
2. (1, -5)
3. $\left(4, \frac{7}{2}\right)$
4. (11, -9)
5. $\left(-\frac{63}{2}, -\frac{57}{2}\right)$
6. $x = 12, y = -18$
7. $\sqrt{34}$
8. 5
9. 10
10. 5
11. $2\sqrt{2}$
12. $a = 6$
13. $c = 5$
14. $b = 5\sqrt{3}$
15. $a = 7$
16. $P = 24$ units, $A = 24$ square units, The perimeter and area are equal.
17. About 3:37 p.m.
18. a) $\left(\frac{c}{2}, \frac{b}{2}\right)$, b) Hint: Show that line segments $MX = MY = MZ$.
19. Yes. Hint: Find the distances of line segments PQ, QR and PR. Show that distance PQ + distance QR = distance PR.
20. a) $A'(-2, 2)$, $B'(-5, 1)$, b) Trapezoid, c) 7 square units
21. a) right triangle, b) $M'(2, 2)$, $N'(2, 3)$, $O'(4, 2)$, c) $\frac{3-a}{2-a}\sqrt{\frac{2a^2-6a+9}{2a^2-2a+5}}$

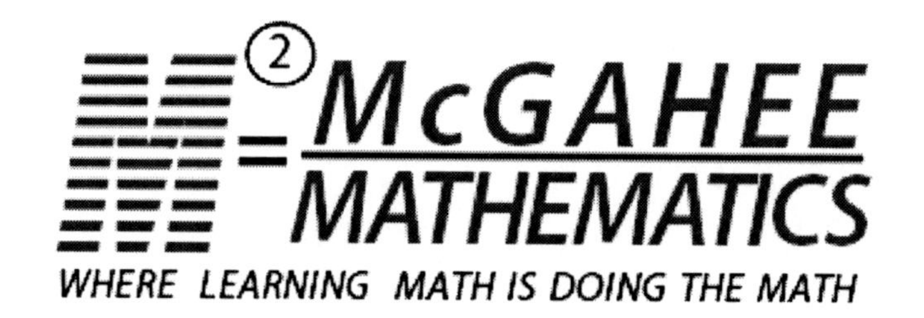

PROPERTIES OF LOGARITHMS

QUESTIONS **ANSWERS**

Evaluate each of the following logarithms.

1. $\log_2 4$ ______
2. $\log_3 1$ ______
3. $\log 10$ ______
4. $\log_5 \frac{1}{125}$ ______
5. $\log_4 \frac{1}{16}$ ______
6. $\log 0$ ______

Simplify the logarithmic expressions.

7. $\log_2 (8 \bullet 16)$ ______
8. $\log_3 \left(\frac{1}{27}\right)$ ______
9. $\log \sqrt[4]{10}$ ______
10. $\log_4 256^7$ ______

Write each of the following expressions as a single logarithm.

11. $\log_3 x + \log_3 y$ ______
12. $2\log_4 w - \log_4 z$ ______
13. $\log a + 2\log b - 3\log c$ ______
14. $\log_2 m^2 - \log_2 n^2 + \log_2 n - \log_2 m$ ______

Determine if the statement is true or false. If it's false, correct the statement to make it true.

15. It is possible to take the logarithm of a negative number. True OR False
16. If $0 < x < 1$, then $log(x) < 0$. True OR False
17. If $\log_b x = M$ and $\log_b y = N$, then $\log_b(x^p y^q) = pM - qN$. True OR False

Simplify each of the following abstract logarithmic expressions.

18. $\log_a \sqrt{ax} - \log_a x + \log_a \sqrt{a}$ ____________________
19. $\log_a x^m y^n - m\log_a x + n\log_a \sqrt[n]{y}$ ____________________

CHALLENGE

20. The natural logarithm $\ln x = \log_e x$ has the natural base e. Use the properties of logarithms to show that

a) $\ln e = 1$ ____________________
b) $e^{\ln x} = x$ ____________________
c) $\ln x^{-1} = -\ln x$ ____________________

21. Suppose a and b are positive integers greater than 1 and x is a positive real number. Show that $\log_a b \log_b a = 1$.

____

SOLUTIONS

1. 2
2. 0
3. 1
4. -3
5. -2
6. Undefined
7. 7
8. -3
9. $\frac{1}{4}$
10. 28
11. $\log_3(xy)$
12. $\log_4\left(\frac{w^2}{z}\right)$
13. $\log\left(\frac{ab^2}{c^3}\right)$
14. $\log_2\left(\frac{m}{n}\right)$
15. False. It is only possible to take the logarithm of a negative number.
16. True.
17. False. $\log_b(x^p y^q) = p\log_b x + q\log_b y$
18. $1 - \frac{1}{2}\log_a x$
19. $(n+1)\log_a y$
20. a) Hint: Let $x = e$.

    b) Hint: Use the definition of logarithm. $\log_b x = y$ if and only if $b^y = x$

    c) Hint: $x^{-1} = \frac{1}{x}$ Use logarithm property $\ln\left(\frac{a}{b}\right) = \ln a - \ln b$.

21. Hint: Use change of base formula. $\log_a x = \frac{\log_b x}{\log_b a}$ and let $x = b$.

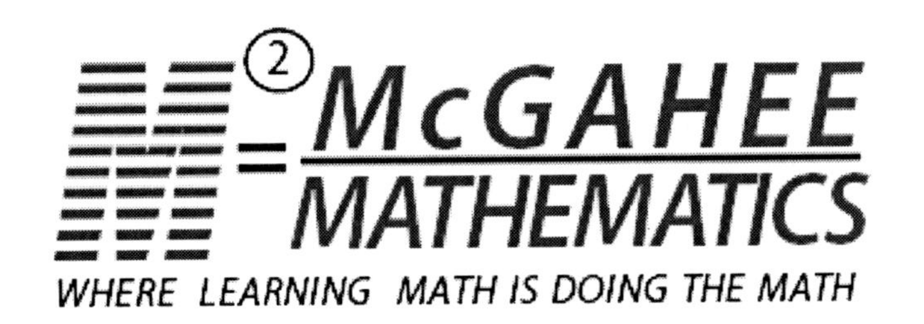

EXPONENTIAL AND LOGARITHMIC EQUATIONS

QUESTIONS **ANSWERS**

Solve each of the following exponential equations.

1. $2^x = 8$ __________
2. $3^y = \frac{1}{243}$ __________
3. $4^{7z-5} = 16$ __________
4. $5^{t+1} = 125^{t-1}$ __________
5. $8^{4r-10} = 32^{20+r}$ __________
6. $10^{a^2} = 100^a$ __________
7. $2^n - 3 = 4$ __________
8. $7^{p+6} = 9^{5-p}$ __________
9. $5e^x + 8 = 23$ __________
10. $e^{2y} - 3e^y + 2 = 0$ __________

Solve each of the following logarithmic equations.

11. $\log_2(x-1) = 4$ __________
12. $\log_3 y^2 = -2$ __________
13. $\log z - \log(z+1) = 1$ __________
14. $\log_4(t+3) + \log_4(t-3) = \log_4 8t$ __________
15. $\ln(2p+3) - \ln p = -5$ __________

Solve each word problem.

16. A bacteria culture has an initial population of 2,500 bacteria. The culture is declining at a rate of 4.1% per hour. About how many hours will it take for the bacteria population to be cut in half?

____

17. An investor deposits an initial amount of $10,000 into a trust fund. The money is compounded quarterly at an interest rate of 5%. How much money will the investor have in the trust fund after 25 years?

____

Solve each abstract exponential/logarithmic equation.

18. $\frac{e^x + a}{e^x - a} = a$ Identify the possible real values for a.

____

19. $\ln(be^{x+b}) = b - x$. Identify the possible real values for b.

____

CHALLENGE

20. Solve $a^{mx+b} = b^{mx+a}$. Write the solution as a quotient of two natural log expressions.

____

21. Solve $\left(\frac{e^{ax} + e^{-ax}}{e^{ax} - e^{-ax}}\right) = a$. Hint: Try substitution $u = e^{ax}$.

____

SOLUTIONS

1. $x = 3$
2. $y = -5$
3. $z = 1$
4. $t = 2$
5. $r = \frac{130}{7}$
6. $a = 0$ or $a = 2$
7. $n = \log_2 7$
8. $p = \frac{5\ln 9 - 6\ln 7}{\ln 63}$
9. $x = \ln 3$
10. $y = \ln 2$ or 0
11. $x = 17$
12. $y = \pm\frac{1}{3}$
13. No solution
14. $t = 9$
15. No solution
16. About 17 hours
17. $34,634.04
18. $x = \ln\left(\frac{a(a+1)}{a-1}\right)$, where $a = (-1,0) \cup (1,\infty)$
19. $x = -\ln\sqrt{b}$, where $b = (0,\infty)$
20. $x = \frac{\ln\left(\frac{b^a}{a^b}\right)}{\ln\left(\frac{a}{b}\right)^m}$
21. $x = \ln\left(\frac{a \pm \sqrt{a^2 - 4a - 4}}{2}\right)^{\frac{1}{a}}$

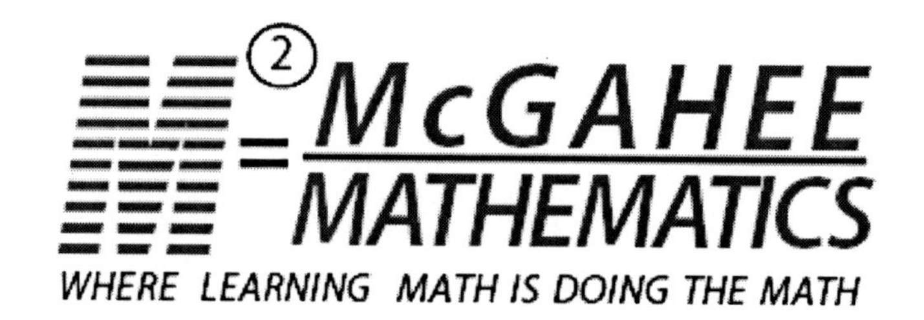

RELATIONS AND FUNCTIONS

QUESTIONS **ANSWERS**

Determine if each relation represents a function. If the relation is a function, find the domain and range.

1. R = {(1, 0), (2, 1), (3, 4), (4, 6), (5, 10)} ____________
2. T = {(-2, -3), (-5, -3), (0, 2), (-5, 4)} ____________
3. W = {(a, 1), (b, 2), (c, 3), (d, 2), (e, 4), (f, 6), (g, 8)} ____________
4. F = {(1, yes), (0, no), (0, yes)} ____________
5. G = {(@, *), ($, *), (&, *), (#, *)} ____________

Determine whether each function is linear or non-linear.

6. $f(x) = -2x + 5$ ____________
7. $f(x) = \sqrt{x+1}$ ____________
8. $f(x) = x^2 - 4x + 8$ ____________
9. $f(x) = -7$ ____________

Perform each of the following function operations and simplify.

Let $f(x) = 3x + 2$ and $g(x) = \frac{1}{x}$.

10. a) $f(5)$ b) $g\left(\frac{1}{2}\right)$ c) $(f+g)(-3)$ a) ________ b) ________ c) ________

11. a) $f(2) + g(-1)$ b) $(f-g)\left(\frac{1}{6}\right)$ c) $(fg)(-4)$ a) ________ b) ________ c) ________

12. Find $(f+g)(x)$, $(f-g)(x)$, $(fg)(x)$, and $\left(\frac{f}{g}\right)(x)$. ________________

Write a function for each of the following situations.

13. The perimeter P of a square with side s. ______________
14. The number of dollars D in q quarters. ______________
15. The volume V of a cube with side x. ______________
16. The number of ounces O in pounds p. ______________
17. The amount of miles M in f feet. ______________

Solve each word problem.

18. A water tank that holds 500 gallons is full to the top and drains at a constant rate of 2.5 gallons per hour. Find a function $V(t)$ that expresses the volume of the tank in t hours. How much water is left in the tank after 1 day? How long will it take for the tank to be completely empty? Express the time in days and hours.

---

19. The height of a stone falling off a 1,200 foot building is modeled by the function $h(t) = -16t^2 + 1{,}200$, where h is the height in feet and t is the time in seconds. What is the height of the stone after 3 seconds? At what time will the stone hit the ground? Round to the nearest tenth of a second.

CHALLENGE

20. Given the function $f(x) = \frac{1}{\sqrt{x}+1}$, find $\frac{f(a+h) - f(a)}{h}$ and evaluate this expression if $h = 0$. This is known as the "difference quotient," which is used in calculus.

---

21. Let $f(x) = mx$ and $g(x) = x^n$, where $m, n \neq 0$. Show that for any numbers a and b, $f(a+b) = f(a) + f(b)$ and $g(ab) = g(a)g(b)$.

SOLUTIONS

1. Yes. Domain = {1, 2, 3, 4, 5}, Range = {0, 1, 4, 6, 10}
2. No.
3. Yes. Domain = {a, b, c, d, e, f, g}, Range = {1, 2, 3, 4, 6}
4. No.
5. Yes. Domain = {@, $, &, #}, Range = {*}
6. Linear
7. Non-linear
8. Non-linear
9. Linear
10. a) 17, b) 2, c) $-\frac{22}{3}$
11. a) 8, b) $-\frac{7}{2}$, c) $\frac{5}{2}$
12. $3x+2-\frac{1}{x}$, $3x+2-\frac{1}{x}$, $3+\frac{2}{x}$, $3x^2+2x$
13. $P(s)=4s$
14. $D(q)=\frac{q}{4}$
15. $V(x)=x^3$
16. $O(p)=16p$
17. $M(f)=\frac{f}{5{,}280}$
18. $V(t)=500-2.5t$, 440 gallons, 8 days and 8 hours
19. 1,056 feet, 8.7 seconds
20. $-\frac{1}{2\sqrt{a}(\sqrt{a}+1)^2}$
21. $f(a+b)=m(a+b)=ma+mb=f(a)+f(b)$

    $g(ab)=(ab)^n=a^nb^n=g(a)g(b)$

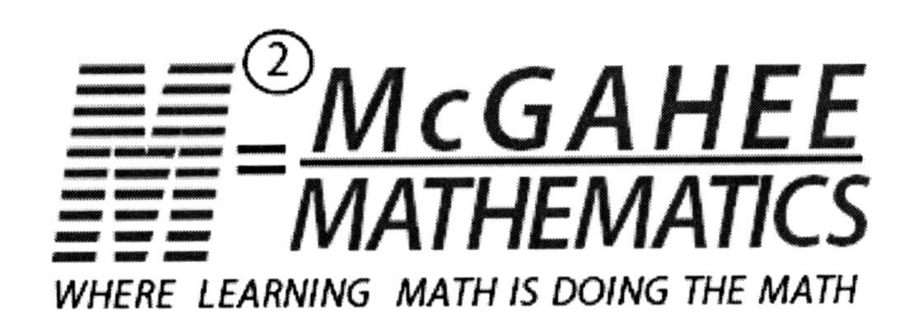

NON-LINEAR FUNCTIONS

QUESTIONS **ANSWERS**

Identify the type of each non-linear function. State the domain and range in interval notation.

1. $f(x) = \sqrt{x-3} + 2$ ____________________
2. $f(x) = -x^2 + 4x - 7$ ____________________
3. $f(x) = |5x - 10|$ ____________________
4. $f(x) = \dfrac{3x+6}{8-x}$ ____________________
5. $f(x) = 2^{x-1} + 1$ ____________________
6. $f(x) = \log(4x + 20)$ ____________________

For each quadratic function, find the vertex, axis of symmetry, maximum or minimum value of the function. Graph the function.

7. $f(x) = 2x^2 + 4x$ ____________________
8. $f(x) = -3x^2 - 9x + 5$ ____________________
9. $f(x) = x^2 - 8$ ____________________

10. Find a quadratic function (in vertex form) of a parabola that has a vertex located at (-1, -3) and passes through (1, -5).

____

Graph each of the following non-linear functions. State the domain, range, and any asymptotes that may occur.

11. $f(x) = \sqrt{2x-4}$ ____________________

12. $f(x) = \dfrac{x^2-9}{x^3+2x^2-3x}$ ____________________

13. $f(x) = e^x$ ____________________

14. $f(x) = -\log x + 5$ ____________________

15. $f(x) = 4|x+2|-6$ ____________________

A function is odd if $f(-x) = -f(x)$ and even if $f(-x) = f(x)$. Determine if each function is even, odd, or neither.

16. $f(x) = x^2 + x^4$ ____________________

17. $f(x) = \dfrac{2}{x}$ ____________________

18. $f(x) = 1 + \sqrt[3]{x} - x$ ____________________

19. If two functions f and g are both odd, then show that $f + g$ is also odd.

Hint: Use the definition of even and odd functions for questions 16-18.

CHALLENGE

20. Suppose that $f(x) = ax^2 + bx + c$ and that $a > 0$. Find the vertex in terms of a, b, and c. Determine if $f(x)$ contains a maximum or a minimum value.

____

21. The period of a pendulum can be expressed by the function:

$T(l) = 2\pi\sqrt{\dfrac{l}{g}}$, where l is the length of the pendulum and g is the gravitational constant. What value of l gives the pendulum a period of 2π? What value of l doubles the period? ____________________

SOLUTIONS

1. Radical Function, Domain = $[3,\infty)$, Range = $[2,\infty)$
2. Quadratic Function, Domain = $(-\infty,\infty)$, Range = $(-\infty,-3]$
3. Absolute Value Function, Domain = $(-\infty,\infty)$, Range = $[0,\infty)$
4. Rational Function, Domain = $(-\infty,8)\cup(8,\infty)$, Range = $(-\infty,-3)\cup(-3,\infty)$
5. Exponential Function, Domain = $(-\infty,\infty)$, Range = $(1,\infty)$
6. Logarithmic Function, Domain = $(-5,\infty)$, Range = $(-\infty,\infty)$
7. Vertex = (-1, -2), Axis of symmetry: x = -1, Minimum value = -2

8. Vertex = $\left(-\frac{3}{2},\frac{47}{4}\right)$, Axis of symmetry: x = $-\frac{3}{2}$, Maximum value = $\frac{47}{4}$
9. Vertex = (0, -8), Axis of symmetry: x = 0, Minimum value = -8

GRAPHS FOR QUESTIONS 7-9

7

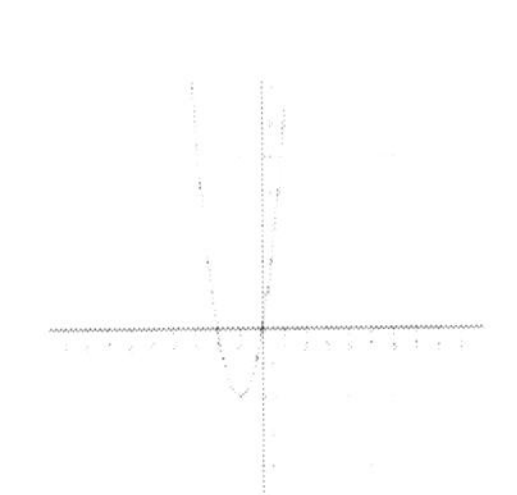

8

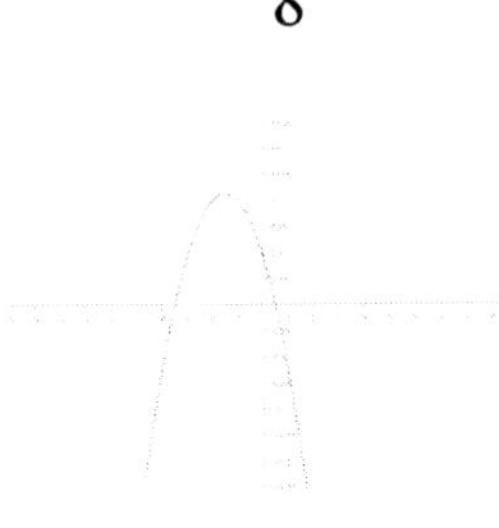

9

10. $f(x) = -\frac{1}{2}(x+1)^2 - 3$

11. Domain = $[2,\infty)$, Range = $[0,\infty)$

12. Domain = $(-\infty,-3)\cup(-3,0)\cup(0,1)\cup(1,\infty)$, Range = $(-\infty,0)\cup(0,\infty)$,
Horizontal Asymptote: y = 0, Vertical Asympotes: x = -3, x = 0, x = 1

13. Domain = $(-\infty,\infty)$, Range = $(0,\infty)$, Horizontal Asymptote: y = 0

14. Domain = $(0,\infty)$, Range = $(-\infty,\infty)$, Vertical Asymptote: x = 0

15. Domain = $(-\infty,\infty)$, Range = $[0,\infty)$

16. Even

17. Odd

18. Neither

GRAPHS FOR QUESTIONS 11-15

11.

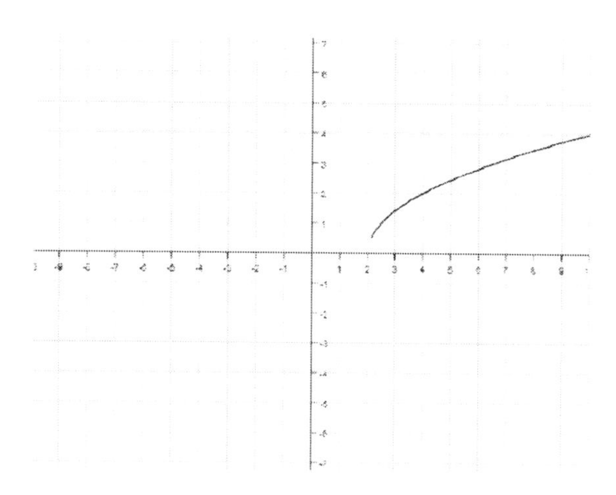

12.

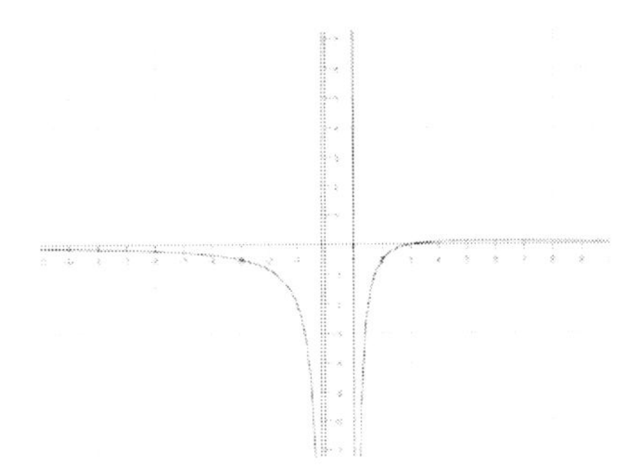

13.

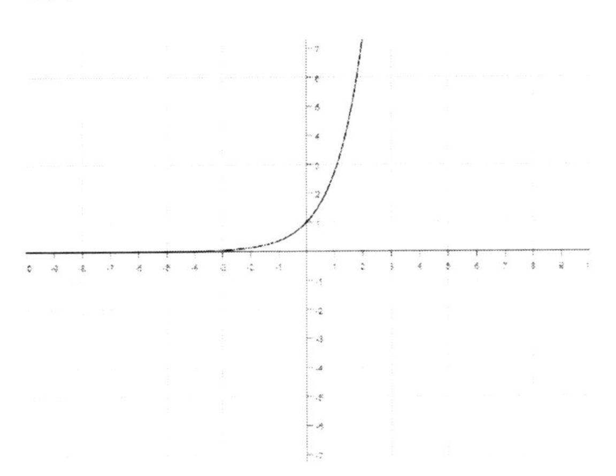

14.

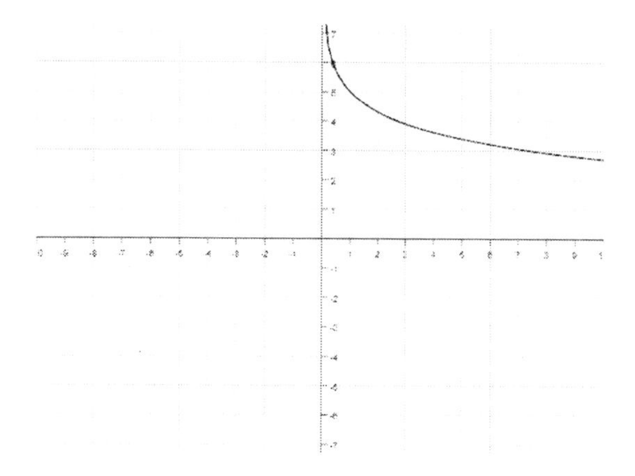

15.

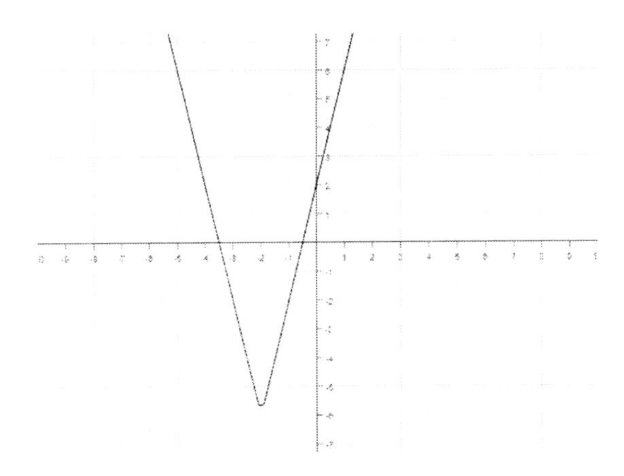

16. Even

17. Odd

18. Neither

19. Let f and g be two odd functions. Then $f(-x) = -f(x)$ and $g(-x) = -g(x)$. This implies that $(f + g)(-x) = f(-x) + g(-x) = -f(x) - g(x) = -(f(x) + g(x)) = -(f + g)(x)$. So, $f + g$ is an odd function.

20. Vertex $= \left(-\frac{b}{2a}, \frac{4ac - b^2}{4a}\right)$, Since $a > 0$, the minimum value is $\frac{4ac - b^2}{4a}$.

21. $l = g$, $l = 4g$

COMPOSITE AND INVERSE FUNCTIONS

QUESTIONS **ANSWERS**

Let $f(x) = x^2 + 4$, $g(x) = \sqrt{x-1}$, and $h(x) = \frac{3}{x}$. Perform each of the following function compositions.

1. $g(f(1))$ __________
2. $f(g(5))$ __________
3. $h(f(-2))$ __________
4. $g(h(3))$ __________

5. a) Find $g(f(x))$, $h(g(x))$, and $f(h(x))$. __________

   b) What is the domain and range of $g(f(x))$? __________

Consider the following table of values. Perform each of the following function compositions.

x	f(x)	g(x)	h(x)
0	1	2	0
1	0	1	2
2	2	0	1

6. $f(g(0))$ __________
7. $g(f(1))$ __________
8. $h(g(2))$ __________

9. $g(h(f(0)))$ ____________

10. $h(f(g(1)))$ ____________

11. Find the value(s) of x such that $h(f(x)) = f(h(x))$. If there are no values that satisfy this equation, write NONE. ____________

Determine if the function is one-to-one. If it is, find its inverse.

12. $f(x) = 2x + 5$ ____________

13. $f(x) = \dfrac{x-1}{x+1}$ ____________

14. $f(x) = -3x^2 + 4x - 7$ ____________

15. $f(x) = \log_2(x+8)$ ____________

16. $f(x) = e^{-6x} + 10$ ____________

17. If $f(x) = x^3 - 2$, and $g(x) = \sqrt[3]{x+2}$, then

a) Show that $f(x)$ and $g(x)$ are both one-to-one functions.

____

b) Show that $f(g(x)) = g(f(x))$ and interpret this result.

____

For True or False questions on #18, if the statement is true, give a reason for why it is true. If the statement is false, give an example to show why the statement is false.

18. True or False: For any functions $f(x)$ and $g(x)$, $f(g(x)) = g(f(x))$. ______

True or False: For any functions $f(x)$ and $f^{-1}(x)$, $f(f^{-1}(x)) = f^{-1}(f(x)) = x$. ______

True or False: For any functions $f(x)$, $g(x)$, and $h(x)$,
$(f \circ (g \circ h))(x) = ((f \circ g) \circ h)(x)$. ______

True or False: For any functions $f(x)$ and $g(x)$,
$(f \circ g)^{-1}(x) = (f^{-1} \circ g^{-1})(x)$. ______

19. If $h(x) = g(f(x))$ such that $h(x) = 4x^2 - 4x + 4$ and $f(x) = 2x - 1$, what is $g(x)$? Find the value of $f^{-1}(g(h(1)))$.

____

CHALLENGE

20. Let $f(x) = mx$. Find $(f \circ f \circ f)(x)$. What happens if you compose $f(x)$ n number of times?

____

21. Let $h = g \circ f$. If g is even and f is odd, then what is h?

____

SOLUTIONS

1. 2
2. 8
3. $\frac{3}{8}$
4. 0
5. a) $g(f(x)) = \sqrt{x^2+3}$, $h(g(x)) = \frac{3}{\sqrt{x-1}}$, $f(h(x)) = \frac{4x^2+9}{x^2}$

   b) Domain = $(-\infty,\infty)$, Range = $[\sqrt{3},\infty)$
6. 2
7. 2
8. 0
9. 0
10. 0
11. NONE
12. $f^{-1}(x) = \frac{x-5}{2}$
13. $f^{-1}(x) = \frac{x+1}{1-x}$
14. Not one-to-one
15. $f^{-1}(x) = 2^x - 8$
16. $f^{-1}(x) = -\frac{1}{6}\ln(x-10)$
17. a) Hint: Let $f(a) = f(b)$. Show that $a = b$. Let $g(a) = g(b)$. Show that $a = b$. b) Show that $f(g(x)) = g(f(x)) = x$. The functions f and g are inverses of each other.
18. a) False. Let $f(x) = x+1$ and $g(x) = \sqrt{x}$. Many other examples will suffice. b) True. c) True. d) False. Let $f(x) = x-1$ and $g(x) = x^3$. Find the inverses of f and g, and show that the compositions are not equal.
19. $g(x) = x^2+3$, $f^{-1}(g(h(1))) = 10$
20. $(f \circ f \circ f)(x) = m^3x$, Composing $f(x)$ n times will give you $f(x) = m^nx$.
21. Even

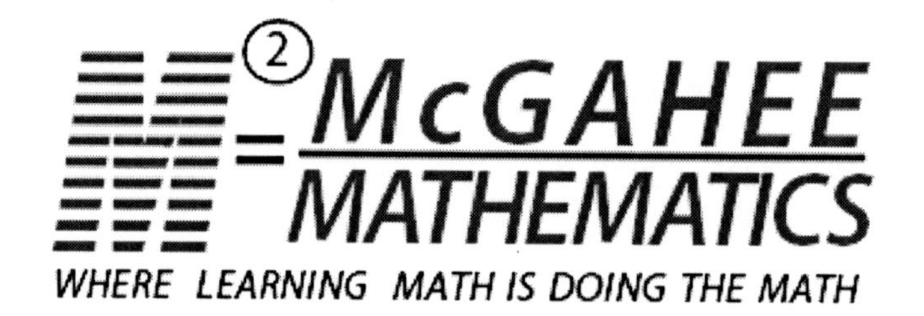

MATRICES, DETERMINANTS, AND CRAMER'S RULE

QUESTIONS **ANSWERS**

Let $A = \begin{bmatrix} 1 & -2 \\ 3 & 0 \end{bmatrix}$, $B = \begin{bmatrix} -4 & 5 \\ 8 & -9 \end{bmatrix}$, $C = \begin{bmatrix} 3 & -6 & 12 \\ -1 & 0 & -7 \end{bmatrix}$, and $D = \begin{bmatrix} -5 \\ 1 \\ 0 \end{bmatrix}$.

Compute each of the following matrix operations (if possible).

1. $A + B$ ______________
2. BA ______________
3. $C - D$ ______________
4. AC ______________
5. CD ______________
6. $B + D$ ______________
7. AD ______________
8. $-3D$ ______________
9. $2A + 3B$ ______________
10. C^2 ______________

11. Compute det(A) and det(B). ______________

12. Find A^{-1} and B^{-1}. ______________

13. Does the matrix $M = \begin{bmatrix} 1 & 0 & -1 \\ -4 & 2 & 0 \\ 0 & 0 & 0 \end{bmatrix}$ have an inverse? Explain.

---

14. Solve the system $\begin{array}{c} 2x - y = 3 \\ x + 5y = -4 \end{array}$ using Cramer's Rule.

---

15. Solve the system $\begin{array}{r} x + y - z = -1 \\ -3x + 2y + z = 17 \\ x - 4y + 2z = -12 \end{array}$ using Cramer's Rule.

---

16. True or False: If matrix A has dimension m x n and matrix B has dimension n x p, then matrix AB has dimension m x p. True OR False

17. True or False: If an n x n matrix has a zero determinant, then it has an inverse. True OR False

18. True or False: Two or more matrices can be added or subtracted if and only if they have the same dimension. True OR False

19. If A and B are both 2 x 2 matrices, then is $A + B = B + A$ true? How about $AB = BA$? If not, provide an example to show why the equation is false.

CHALLENGE

20. If $A = \begin{bmatrix} a & -b \\ b & a \end{bmatrix}$, then find A^{-1}. Verify your answer by showing that

$AA^{-1} = A^{-1}A = I$, where $I = \begin{bmatrix} 1 & 0 \\ 0 & 1 \end{bmatrix}$ is the 2 x 2 identity matrix.

____

21. If $A = \begin{bmatrix} 1 & 0 \\ 1 & 1 \end{bmatrix}$, find A^n. Hint: Start finding A^2, A^3, etc. until you see a pattern.

____

SOLUTIONS

1. $A+B=\begin{bmatrix} -3 & 3 \\ 11 & -9 \end{bmatrix}$
2. $BA=\begin{bmatrix} 11 & 8 \\ -19 & -16 \end{bmatrix}$
3. Impossible
4. $AC=\begin{bmatrix} 5 & -6 & 26 \\ 9 & -18 & 36 \end{bmatrix}$
5. $CD=\begin{bmatrix} -21 \\ 5 \end{bmatrix}$
6. Impossible
7. Impossible
8. $-3D=\begin{bmatrix} 15 \\ -3 \\ 0 \end{bmatrix}$
9. $2A+3B=\begin{bmatrix} -10 & 11 \\ 30 & -27 \end{bmatrix}$
10. Impossible
11. det(A) = 6, det(B) = -4
12. $A^{-1}=\begin{bmatrix} 0 & \frac{1}{3} \\ -\frac{1}{2} & \frac{1}{6} \end{bmatrix}$, $B^{-1}=\begin{bmatrix} \frac{9}{4} & \frac{5}{4} \\ 2 & 1 \end{bmatrix}$
13. No, since det(M) = 0.
14. (1, -1)
15. (-2, 4, 3)
16. True.
17. False. Matrix with a zero determinant does not have an inverse.
18. True.
19. True. False. E.g. Let $A=\begin{bmatrix} 1 & 0 \\ -1 & 1 \end{bmatrix}$ and $B=\begin{bmatrix} 0 & 1 \\ 1 & -1 \end{bmatrix}$. Then

    $AB=\begin{bmatrix} 0 & 1 \\ 1 & -2 \end{bmatrix}$ and $BA=\begin{bmatrix} -1 & 1 \\ 2 & -1 \end{bmatrix}$. So, $AB \neq BA$.

20. $A^{-1} = \begin{bmatrix} \frac{a}{a^2+b^2} & \frac{b}{a^2+b^2} \\ -\frac{b}{a^2+b^2} & \frac{a}{a^2+b^2} \end{bmatrix}$

21. $A^n = \begin{bmatrix} 1 & 0 \\ n & 1 \end{bmatrix}$

FINAL EXAM

Directions: Answer each of the 20 questions completely on your own. Show all work to receive full credit.

When you are finished, email your final exam to Ben McGahee at bmcgahee@mcgaheemathematics.com to have it graded.

Good Luck! ☺

1. Solve. $6(x-8)+7=-2x-14$ __________________

2. Solve. $|4y+5|-10=23$ __________________

3. Solve. $\frac{2}{3}(9z-27)\le\frac{z}{6}$ __________________

4. Write $3x+6y=12$ in slope-intercept form. __________________
   Graph the line.

5. Find the value of k such that slope of the line passing through the points (1, -2) and (3, k) is -1. __________________

6. Solve the system of equations by any method. __________________

   $3s+4t=7$
   $s-2t=-5$

7. Simplify the polynomial. $8x^3 + 11x^2 - (7x^3 - 14x^2 + 3)$ ____________

8. Multiply. $(m+n)^2(m-n)$ ____________

9. Divide. $\dfrac{20p^4 - 10p^2 + 5p - 15}{5p - 5}$ ____________

10. Factor Completely. $ax^2 + bx - a^2x - ab$ ____________

11. Factor Completely. $9b^4 - 36$ ____________

12. Write the complex number in $a + bi$ form.

$\dfrac{5+7i}{2-4i}$ ____________

13. Solve the quadratic equation. $2x^2 + 6x - 8 = 0$ ____________

14. Solve the rational inequality. $\dfrac{1}{t} - \dfrac{1}{t+1} > \dfrac{1}{t^2 + t}$ ____________

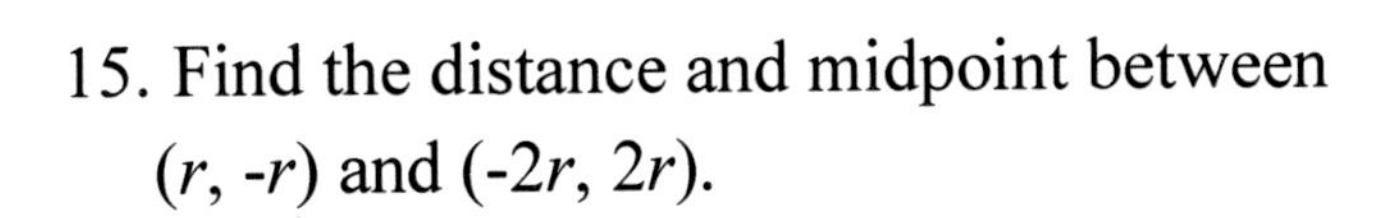

15. Find the distance and midpoint between $(r, -r)$ and $(-2r, 2r)$.

16. Find the domain and range of $f(x) = \dfrac{x-1}{x^2-4}$.

Graph the function. Label any asymptotes.

17. If $f(x) = 2e^{5x+5} - 6$, find $f^{-1}(-4)$.

18. Solve the logarithmic equation.

$$\log_2(x+a) - \log_2(a-x) = -1$$

19. Find AB if $A = \begin{bmatrix} 1 & 2 \\ 3 & 4 \end{bmatrix}$ and $B = \begin{bmatrix} -2 & 0 & 6 \\ 1 & -5 & 7 \end{bmatrix}$

20. Find A^{-1} if $A^{-1} = \begin{bmatrix} 1 & 0 & -2 \\ 0 & -1 & 0 \\ 2 & 1 & 1 \end{bmatrix}$

FORMULA REFRENCE

Commutative Property of Addition/Multiplication: $a+b=b+a,\ ab=ba$

Associative Property of Addition/Multiplication:

$a+(b+c)=(a+b)+c,\ a(bc)=(ab)c$

Distributive Property: $a(b+c)=ab+ac$

Law of Exponents:

Multiply/Add: $x^m x^n = x^{m+n}$

Divide/Subtract: $\frac{x^m}{x^n} = x^{m-n}$

Zero Power: $x^0 = 1, x \neq 0$

Power to a Power: $(x^m)^n = x^{mn}$

Multiplication with a Power: $(xy)^n = x^n y^n$

Division with a Power: $\left(\frac{x}{y}\right)^n = \frac{x^n}{y^n}, y \neq 0$

Slope of a Line: $m = \frac{y_2 - y_1}{x_2 - x_1}$

Slope-Intercept Form of a Line: $y = mx + b$

Difference of Two Squares: $a^2 - b^2 = (a+b)(a-b)$

Sum/Difference of Cubes: $(a \pm b)(a^2 \mp ab + b^2)$

Quadratic Formula: $x = \frac{-b \pm \sqrt{b^2 - 4ac}}{2a}, a \neq 0$

Vertex Formula: $x = -\frac{b}{2a}, y = \frac{4ac - b^2}{4a}$

Quadratic Function (Vertex Form): $f(x) = a(x-h)^2 + k$

Pythagorean Theorem: $a^2 + b^2 = c^2$

Radical to Exponent Formula: $x^{\frac{m}{n}} = \sqrt[n]{x^m} = \left(\sqrt[n]{x}\right)^m$

Multiplying Radicals: $\sqrt[n]{xy} = \sqrt[n]{x}\sqrt[n]{y}$

Dividing Radicals: $\sqrt[n]{\frac{x}{y}} = \frac{\sqrt[n]{x}}{\sqrt[n]{y}}, y \neq 0$

Distance Formula: $d = \sqrt{(x_1 - x_2)^2 + (y_1 - y_2)^2}$

Midpoint Formula: $\left(\frac{x_1 + x_2}{2}, \frac{y_1 + y_2}{2}\right)$

Product of Complex Number and its Conjugate: $(a+bi)(a-bi) = a^2 + b^2$

Logarithm Formulas:

Product-Sum Rule: $\log_b(xy) = \log_b x + \log_b y$

Quotient-Difference Rule: $\log_b\left(\frac{x}{y}\right) = \log_b x - \log_b y$

Power Rule: $\log_b x^p = p \log_b x$

Change of Base Formula: $\log_b x = \frac{\log_a x}{\log_a b}$

Function Operations:

$$(f+g)(x)=f(x)+g(x)$$
$$(f-g)(x)=f(x)-g(x)$$
$$(fg)(x)=f(x)g(x)$$
$$\left(\frac{f}{g}\right)(x)=\frac{f(x)}{g(x)},\ g(x)\neq 0$$

Composition of Functions:

$$(f\circ g)(x)=f(g(x))$$
$$(g\circ f)(x)=g(f(x))$$
$$(f\circ f^{-1})(x)=(f^{-1}\circ f)(x)=x$$
$$(f\circ(g\circ h))(x)=((f\circ g)\circ h)(x)$$
$$(f\circ g)^{-1}(x)=(g^{-1}\circ f^{-1})(x)$$

Composition of e^x and $\ln x$: $e^{\ln x}=\ln e^x=x$

Determinants:

2 x 2 Matrix:

$$\det\left(\begin{bmatrix} a & b \\ c & d \end{bmatrix}\right)=ad-bc$$

3 x 3 Matrix:

$$\det\left(\begin{bmatrix} a & b & c \\ d & e & f \\ g & h & i \end{bmatrix}\right)=a(ei-fh)-b(di-fg)+c(dh-eg)$$

Product of Matrix and its Inverse: $AA^{-1}=A^{-1}A=I$

CPSIA information can be obtained at www.ICGtesting.com
Printed in the USA
BVOW06s2203070714

358424BV00006B/50/P